Selected Titles in This Series

Generalized Minkowski Content, Spectrum of Fractal Drums, Fractal Strings and the Riemann-Zeta-Function

$\mathbf{M}$EMOIRS

of the
American Mathematical Society

Number 608

Generalized Minkowski Content, Spectrum of Fractal Drums, Fractal Strings and the Riemann Zeta-Function

Christina Q. He
Michel L. Lapidus

May 1997 • Volume 127 • Number 608 (end of volume) • ISSN 0065-9266

American Mathematical Society
Providence, Rhode Island

1991 *Mathematics Subject Classification.*
Primary 35P20, 11M06, 28A12, 28A80, 34B24, 58F19;
Secondary 11M26, 26B15, 28A75, 35J20, 47A75, 78A40, 78A45.

Library of Congress Cataloging-in-Publication Data

He, Christina Q., 1970–
    Generalized Minkowski content, spectrum of fractal drums, fractal strings, and the Riemann zeta-function / Christina Q. He, Michel L. Lapidus.
       p. cm. — (Memoirs of the American Mathematical Society, ISSN 0065-9266 ; no. 608)
    "May 1997, volume 127, number 608 (end of volume)."
    Includes bibliographical references.
    ISBN 0-8218-0597-5 (alk. paper)
    1. Differential equations, Partial—Numerical solutions. 2. Spectral theory (Mathematics) 3. Functions, Zeta. 4. Fractals. I. Lapidus, Michel L. (Michel Laurent), 1956– . II. Title. III. Series.
QA3.A57 no. 608
[QA377]
510 s—dc21
[515′.353]
                                              97-422
                                               CIP

Memoirs of the American Mathematical Society

This journal is devoted entirely to research in pure and applied mathematics.

Subscription information. The 1997 subscription begins with number 595 and consists of six mailings, each containing one or more numbers. Subscription prices for 1997 are $414 list, $331 institutional member. A late charge of 10% of the subscription price will be imposed on orders received from nonmembers after January 1 of the subscription year. Subscribers outside the United States and India must pay a postage surcharge of $30; subscribers in India must pay a postage surcharge of $43. Expedited delivery to destinations in North America $35; elsewhere $110. Each number may be ordered separately; *please specify number* when ordering an individual number. For prices and titles of recently released numbers, see the New Publications sections of the *Notices of the American Mathematical Society.*

Back number information. For back issues see the *AMS Catalog of Publications.*

Subscriptions and orders should be addressed to the American Mathematical Society, P. O. Box 5904, Boston, MA 02206-5904. *All orders must be accompanied by payment.* Other correspondence should be addressed to Box 6248, Providence, RI 02940-6248.

Memoirs of the American Mathematical Society is published bimonthly (each volume consisting usually of more than one number) by the American Mathematical Society at 201 Charles Street, Providence, RI 02904-2294. Periodicals postage paid at Providence, RI. Postmaster: Send address changes to Memoirs, American Mathematical Society, P. O. Box 6248, Providence, RI 02940-6248.

Contents

Abstract

This paper provides a detailed study of the effect of non power-like irregularities of (the geometry of) the fractal boundary on the spectrum of 'fractal drums' (and especially, of 'fractal strings').

In earlier work [La1]—devoted to the standard case of power-type irregularities—the second author obtained a partial resolution of the Weyl-Berry conjecture for the vibrations of 'fractal drums' (i.e., 'drums with fractal boundaries'); he thereby obtained sharp error estimates for the asymptotics of the eigenvalue distribution of the Dirichlet (or Neumann) Laplacian on an open subset Ω of $\mathbf{R}^n$ with finite volume and very irregular ('fractal') boundary $\Gamma = \partial\Omega$. Further, when $n = 1$, Lapidus and Pomerance [LaPo1,2] made a detailed study of the corresponding direct spectral problem for the vibrations of 'fractal strings' (i.e., one-dimensional 'fractal drums') and established in the process some unexpected connections with the Riemann zeta-function $\zeta = \zeta(s)$ in the 'critical interval' $0 < s < 1$. Later on (still when $n = 1$), using the oscillatory phenomena associated with the complex zeros of ζ in the 'critical strip' $0 < Re\ s < 1$, Lapidus and Maier [LaMa1,2] obtained a new characterization of the Riemann hypothesis by means of an associated inverse spectral problem.

In this memoir, we will extend most of these results by using, in particular, the notion of generalized Minkowski content which is defined through some suitable 'gauge functions' other than the power functions. [This content is used to measure the irregularity (or 'fractality') of the boundary $\Gamma = \partial\Omega$ by evaluating the volume of its small tubular neighborhoods.] In the situation when the power function is not the natural 'gauge function', this will enable us to obtain more precise estimates, with a broader potential range of applications than in the above papers.

Key words and phrases. Generalized Minkowski content, gauge functions, Minkowski dimension, open sets with fractal boundaries, Dirichlet Laplacian, spectrum of fractal drums and strings, asymptotics of eigenvalues, error estimates, inverse spectral problem, Riemann zeta-function, critical zeros.

Acknowledgements

The second author (Michel L. Lapidus) was visiting the *Institut des Hautes Etudes Scientifiques* (IHES) in Bures-sur-Yvette, France, during parts of 1994 while this paper was completed. He would like to thank the IHES for its warm hospitality. He would especially like to thank Professors Alain Connes, Jean-Pierre Bourguignon and Dennis Sullivan for their kind welcome at the institute. He is also grateful to the National Science Foundation for its continued support of his research.

This work was presented in particular by the first author (Christina Q. He) at the Conference on "Wavelets and Fractals" in Pittsburgh in May 1994 as well as at the Workshop of the Association of Women in Mathematics (AWM) held in conjunction with the American Mathematical Society (AMS) Annual Meeting in San Francisco in January 1995.

It was also presented by the second author in a plenary talk at the aforementioned conference in Pittsburgh and during a research course given at the Summer School on "Progress in Inverse Spectral Geometry" held in Stockholm, Sweden, in June 1994, as well as in a plenary lecture at the Symposium on "Fractal Geometry and Self-Similar Phenomena" held in Curaçao, Dutch Antilles, in February 1995 in honor of Benoît Mandelbrot's 70th birthday.

1   Introduction

Let Ω be an open set in $\mathbf{R}^n$ $(n \geq 1)$, with boundary $\Gamma = \partial\Omega$. We assume that Ω is nonempty and with finite volume. Consider the following eigenvalue problem:

$$(P) \qquad \begin{cases} -\Delta u &= \lambda u \quad \text{in } \Omega \\ u &= 0 \quad \text{on } \Gamma, \end{cases}$$

where $\Delta = \sum_{k=1}^{n} \partial^2/\partial x_k^2$ denotes the Dirichlet Laplacian on Ω. In this general setting, the problem (P) is interpreted in the variational sense. More precisely, the scalar λ is said to be an eigenvalue of (P) if there exists $u \neq 0$ in $H_0^1(\Omega)$ (the closure of $C_0^\infty(\Omega)$, the space of smooth functions with compact support contained in Ω, in the Sobolev space $H^1(\Omega)$) that is a *weak* (distributional) solution of (P); i.e., for all $v \in H_0^1(\Omega)$,

$$\int_\Omega \nabla u \cdot \nabla v = \lambda \int_\Omega u \cdot v.$$

As is well known, the spectrum of (P) is discrete and consists of a sequence $(\lambda_k)_{k=1}^\infty$ of eigenvalues (with finite multiplicity) written in increasing order according to their multiplicity:

$$0 < \lambda_1 \leq \lambda_2 \leq \cdots \leq \lambda_k \leq \cdots, \quad \text{with } \lambda_k \to \infty \text{ as } k \to \infty.$$

Let $N(\lambda)$ denote the eigenvalue counting function of (P); that is, for $\lambda > 0$,

$$N(\lambda) = \#\{k \geq 1 : \lambda_k \leq \lambda\}.$$

Generalizing H. Weyl's classical theorem [We], Birmann and Solomjak showed in [BiSo] that

---

Received by the editors in February, 1995.

Research (of M. L. Lapidus) partially supported by the National Science Foundation under grant DMS-9207098.

1

$$N(\lambda) = \varphi(\lambda)(1 + o(1)), \text{ as } \lambda \to \infty, \tag{1.1}$$

where the "Weyl term" is given by

$$\varphi(\lambda) = (2\pi)^{-n}\mathcal{B}_n|\Omega|_n\lambda^{n/2}. \tag{1.2}$$

Also, see the work of Métivier in [Mt] for related results. Here, $|A|_n$ denotes the n-dimensional Lebesgue measure of $A \subset \mathbf{R}^n$ and $\mathcal{B}_n$ is the volume of the unit ball in $\mathbf{R}^n$.

We are interested in the case when the boundary Γ is very irregular; that is, in the 'fractal' case. First let us recall the notions of Minkowski dimension and content.

Definition 1.1 *Given $d > 0$, the d-dimensional upper Minkowski content of $\Gamma = \partial\Omega$ is given by*

$$M^*(d;\Gamma) = \limsup_{\epsilon \to 0^+} \epsilon^{-(n-d)}|\Gamma_\epsilon \cap \Omega|_n, \tag{1.3}$$

where Γ_ϵ, the ϵ-neighborhood of Γ, is the set of $x \in \mathbf{R}^n$ within a distance less than ϵ from Γ. Similarly, we can define the lower Minkowski content $M_*(d;\Gamma)$ by taking the lower limit in (1.3). The Minkowski dimension D of Γ is then defined by

$$D = \inf\{d : M^*(d;\Gamma) < \infty\} = \sup\{d : M^*(d;\Gamma) = \infty\}.$$

Further, we say that Γ is Minkowski measurable if

$$0 < M_*(d;\Gamma) = M^*(d;\Gamma) < \infty \text{ for some } d > 0,$$

and we then call this common value $M(d;\Gamma)$ the Minkowski content of Γ. (In this case, it is clear that $d = D$.)

The Minkowski dimension, also called 'box dimension', was introduced by Bouligand [Bo] for nonintegral values of D. For more information about its properties and its relationships with other 'fractal' dimensions, such as the Hausdorff dimension, see, e.g., [Tr], [La1, §3] and [Fa2, chap. 3].

Following [La1], we will say that Γ is 'fractal' if $D > n - 1$; i.e., if $D \neq n - 1$, the topological dimension of Γ. We stress that *no* assumption of self-similarity, in the sense of [M] or [Fa1,2], is implied here.

In [La1], Lapidus proved the following sharp error estimate.

Theorem 1.2 *Suppose* $\Omega \subset \mathbf{R}^n$ *with finite volume has fractal boundary* Γ *with Minkowski dimension* $D \in (n-1, n)$ *and finite upper Minkowski content* $M^*(D; \Gamma) < \infty$, *then the following remainder estimate holds:*

$$N(\lambda) = \varphi(\lambda) + O(\lambda^{D/2}), \quad as \quad \lambda \to \infty.$$

Further, it is shown in [La1, Examples 5.1 and 5.1′] that this estimate is in general best possible for every $D \in (n-1, n)$, provided that $M^*(D; \Gamma) < \infty$. (If $M^*(D; \Gamma) = \infty$, then [La1, Theorem 2.1] implies that the same estimate holds, but with D replaced by $D + \epsilon$, for any $\epsilon > 0$. We will see in the present paper that the latter result can be improved.)

Theorem 1.2 provides a partial resolution of the 'Weyl-Berry conjecture' for '*fractal drums*' (i.e., for 'drums with fractal boundaries'). (Corresponding pre-Tauberian estimates for the Dirichlet heat kernel were obtained earlier in [BrCa] by Brossard and Carmona who also disproved Berry's original conjecture, expressed in terms of the Hausdorff rather than the Minkowski dimension; also see [LaFl, La2].) Berry's conjecture [Be1,2]—which extends from the 'smooth' to the 'fractal' case Weyl's conjecture [We2]—has significant physical applications to the scattering of waves by rough surfaces or the study of porous media. We refer, e.g., to [La1-4, LaPo2] for a more detailed account of results related to these conjectures.

We now consider the *one-dimensional* case throughout the rest of this section; that is, we assume that $n = 1$. In the terminology of [La1-4, LaPo1-2, LaMa1-2], we therefore work with '*fractal strings*' instead of 'fractal drums'.

In order to state the next theorem precisely, we need to use the classical Riemann zeta-function $\zeta = \zeta(s)$ (see, e.g., [Ti]). Recall that $\zeta(s) = \sum_{j=1}^{\infty} j^{-s}$ for $Re \ s > 1$ and that $\zeta(s)$ has a meromorphic extension to all of $\mathbf{C}$ with a single (simple) pole at $s = 1$. In particular (see [LaPo2, Eq. (2.3), p. 45]),

$$\zeta(s) = \frac{1}{s-1} + \int_1^{\infty} ([t]^{-s} - t^{-s}) dt \quad \text{for} \quad Re \ s > 0. \qquad (1.4)$$

We can next recall two of the main results obtained by Lapidus and Pomerance in [LaPo2].

Theorem 1.3 *If* $\Omega \subset \mathbf{R}$ *has 'fractal' boundary* Γ *which is Minkowski measurable and has Minkowski dimension* $D \in (0, 1)$, *then*

$$N(\lambda) = \varphi(\lambda) - c_{1,D} M(D; \Gamma) \lambda^{D/2} + o(\lambda^{D/2}), \quad as \quad \lambda \to \infty,$$

where the positive constant $c_{1,D}$ is given by

$$c_{1,D} = 2^{D-1}\pi^{-D}(1-D)(-\zeta(D)), \tag{1.5}$$

and $\varphi(\lambda) = \pi^{-1}|\Omega|_1\lambda^{1/2}$.

Theorem 1.3 establishes the 'modified Weyl-Berry conjecture' of [La1, p. 520] when $n = 1$. [The latter is not true when $n \geq 2$; see [FlVa] and [LaPo3] (as well as [La3, §4]).]

Also, the following result is proved in [LaPo2].

Theorem 1.4 *Suppose that $\Omega \subset \mathbf{R}$ has 'fractal' boundary Γ with Minkowski dimension $D \in (0,1)$. Then $0 < M_*(D;\Gamma) \leq M^*(D;\Gamma) < \infty$ if and only if*

$$\varphi(\lambda) - N(\lambda) \asymp \lambda^{D/2}, \quad as \quad \lambda \to \infty.$$

Further, in [LaMa1,2], Lapidus and Maier examined the following *converse* of Theorem 1.3 (still for $n = 1$): let $\Omega \subset \mathbf{R}$ be an open subset with finite length. If for some $D \in (0,1)$,

$$N(\lambda) = \varphi(\lambda) + c\lambda^{D/2} + R(\lambda),$$

where c is a nonzero constant and $R(\lambda) = o(\lambda^{D/2})$ as $\lambda \to \infty$, then $\Gamma = \partial\Omega$ is Minkowski measurable.

More specifically, they considered the 'weak' converse which is the corresponding statement with $R(\lambda) = O(\lambda^{D/2}\ln^{-(1+\epsilon)}\lambda)$, for some $\epsilon > 0$, as $\lambda \to \infty$. They then proved the following theorem.

Theorem 1.5 (characterization of the partial Riemann hypothesis) *Let $D \in (0,1)$. Then the 'weak' converse is true for this value of D if and only if the Riemann zeta-function $\zeta = \zeta(s)$ does not have any zero on the vertical line $\operatorname{Re} s = D$.*

(Note that it follows that this *inverse spectral problem* (i.e., the 'weak' converse) is true for all values of D in the '*critical interval*' $(0,1)$ (other than $1/2$) if and only if the Riemann hypothesis is true; see the corollaries to [LaMa2, Theorems 2.3 and 2.4].)

They established the necessary condition in Theorem 1.5 by proving the contraposition. More precisely, they showed in particular the following theorem, which provides a counterexample to the converse of Theorem 1.3.

Theorem 1.6 *Let $\rho = D + i\nu$ $(0 < D < 1, \nu \in \mathbf{R})$ be a zero of the Riemann zeta-function . Then there exists an open set $\Omega \subset \mathbf{R}$ of finite length and with boundary Γ which is <u>not</u> Minkowski measurable, but such that*

$$N(\lambda) = \varphi(\lambda) + c\lambda^{D/2} + o(\lambda^{D/2}), \quad as \quad \lambda \to \infty,$$

for some nonzero constant c. Moreover, Γ has Minkowski dimension D and in fact, $0 < M_*(D;\Gamma) < M^*(D;\Gamma) < \infty$.

The goal of the present paper is to extend all of these theorems from [La1, LaPo1-2, LaMa1-2] (with the exception of Theorem 1.5), as well as related results in [LaPo2], by using more general 'gauge functions' than the power function in (1.3). This will enable us, in particular, to obtain more precise estimates, with a broader range of applications than in the above papers. In the case of the Hausdorff measure, the use of general 'gauge functions' has allowed more flexibility in many situations (such as the study of Brownian motion or that of exceptional sets in harmonic analysis); see, e.g., [Fa1-2, KaSa, M, Ro, Tr] and the relevant references therein. A similar flexibility is provided in the present case of the generalized Minkowski content.

2 Statement of the Main Results

We now give the generalized definition of h-Minkowski content which will be used throughout this paper. Later on, we will specify the class of 'gauge functions' h needed in different cases: for $n = 1$ in Section 2.1 and for $n \geq 2$ in Section 2.2.

Definition 2.1 *Let $\Omega \subset \mathbf{R}^n$ be an open set with finite volume and boundary $\Gamma = \partial\Omega$. Let $h : (0, \infty) \to (0, \infty)$ be a nondecreasing function. The upper h-Minkowski content of Γ is defined by*

$$M^*(h; \Gamma) = \limsup_{\epsilon \to 0^+} \epsilon^{-n} h(\epsilon) |\Gamma_\epsilon \cap \Omega|_n. \tag{2.1}$$

We define similarly the lower h-Minkowski content $M_*(h; \Gamma)$ by taking the lower limit in (2.1). Further, we say that Γ is h-Minkowski measurable if $0 < M_*(h; \Gamma) = M^*(h; \Gamma) < \infty$, and call this common value $M(h; \Gamma)$ the h-Minkowski content of Γ.

Clearly, the standard definition (recalled in Definition 1.1) just corresponds to the case when $h(x) = x^d$.

2.1 One-dimensional case ($n = 1$)

In the following, we will state the extension of Theorems 1.2—1.4, as well as Theorem 1.6, for $n = 1$.

Let Ω be a nonempty open subset of $\mathbf{R}$ with finite length $|\Omega|_1$ and with boundary $\Gamma = \partial\Omega$. We write Ω as the union of its connected components: $\Omega = \cup_{j=1}^\infty I_j$, where the open intervals I_j are pairwise disjoint and of length l_j. Since $|\Omega|_1 = \sum_{j=1}^\infty l_j < \infty$, we may assume without loss of generality that $l_1 \geq l_2 \geq \cdots \geq l_j \geq \cdots > 0$. We say that $(l_j)_{j=1}^\infty$ is the *sequence associated with Ω*.

Recall that the eigenvalues of the differential operator $-d^2/dy^2$ on the bounded open interval $I := (a, b)$ with Dirichlet boundary conditions at a and b, $u(a) = u(b) = 0$, are $\mu_k = (\pi/l)^2 k^2$ for $k = 1, 2, \ldots$, where $l = b - a$. Let $N(\lambda; I)$ denote the associated eigenvalue counting function. Then if $[\gamma]$ denotes the *integer part* of the real number γ, it follows that for $\lambda > 0$, $N(\lambda; I) = [l\sqrt{\lambda}/\pi]$.

Consequently, we deduce that the counting function associated with Ω is given by

$$N(\lambda) = \sum_{j=1}^{\infty} [l_j x], \quad \text{where } x := \frac{\sqrt{\lambda}}{\pi}. \tag{2.2}$$

In the following, we specify the family of gauge functions that we are going to use in the present case when $n = 1$.

Definition 2.2 *Given $d \in (0, 1)$, let G_d be the class of functions h which satisfy all the following conditions:*

(H1). $h : (0, \infty) \to (0, \infty)$ *is a continuous strictly increasing positive function, and $\lim_{x \to 0^+} h(x) = 0$, $\lim_{x \to \infty} h(x) = \infty$, $\lim_{x \to 0^+} h(x)/x = \infty$.*

(H2). *For any $t > 0$,*

$$\lim_{x \to 0^+} \frac{h(tx)}{h(x)} = t^d,$$

uniformly in t on any compact subset of $(0, \infty)$. (This is a homogeneity condition about $x = 0$.)

(H3). *There exist some constants $\tau \in (0, 1)$, $m > 0$, $0 < x_0 \leq 1$, $0 < t_0 \leq 1$ such that*

$$\frac{h(tx)}{h(x)} \geq m t^\tau, \quad \text{for all } 0 < x < x_0, 0 < t < t_0.$$

Example 2.3 One can check that the functions

$$h(x) = \frac{x^d}{(\ln(\frac{1}{x} + 1))^a} \quad \text{and } h(x) = \frac{x^d}{(\ln(\ln(\frac{1}{x} + 1) + 1))^a}$$

are in G_d for all $d \in (0, 1)$ and $a \geq 0$. See Examples 1 and 2 in the appendix for a detailed proof.

Notation. From now on, given $h \in G_d$, we will always let

$$g(x) := h^{-1}(1/x), \quad f(x) := \frac{1}{h(1/x)}. \tag{2.3}$$

(The existence of the inverse function h^{-1} follows from (H1).)

Let us state our main results when $n = 1$ which generalize the corresponding theorems in Section 1.

Our first result will yield the generalization corresponding to Theorem 1.3.

Theorem 2.4 (a) *Let* $(l_j)_{j=1}^\infty$ *be an arbitrary nonincreasing positive sequence such that for some* $h \in G_d$ *and some constant* $L > 0$, *we have*

$$l_j \sim Lg(j), \quad as \ j \to \infty \tag{2.4}$$

(i.e., $l_j/g(j) \to L$ *as* $j \to \infty$*).*
Then we have

$$\sum_{j=1}^\infty [l_j x] = \left(\sum_{j=1}^\infty l_j \right) x + \zeta(d) L^d f(x) + o(f(x)), \quad as \ j \to \infty. \tag{2.5}$$

(b) *In particular, if the sequence* $(l_j)_{j=1}^\infty$ *associated with* Ω *satisfies hypothesis (2.4), then by letting* $x = \sqrt{\lambda}/\pi$ *in (2.5), we deduce that*

$$N(\lambda) = \varphi(\lambda) + \pi^{-d}\zeta(d)L^d f(\sqrt{\lambda}) + o(f(\sqrt{\lambda})), \quad as \ \lambda \to \infty, \tag{2.6}$$

where, as in (1.2) with $n = 1$, $\varphi(\lambda) = \pi^{-1}|\Omega|_1 \lambda^{1/2} = \pi^{-1}(\sum_{j=1}^\infty l_j)\lambda^{1/2}$.

Theorem 2.5 (characterization of Minkowski measurability) *The sequence* $(l_j)_{j=1}^\infty$ *associated with* Ω *satisfies hypothesis (2.4) if and only if* $\Gamma = \partial\Omega$ *is h-Minkowski measurable. Further, in this case, the h-Minkowski content of* Γ *is given by*

$$M(h; \Gamma) = \frac{2^{1-d}}{1-d} L^d. \tag{2.7}$$

By combining Theorems 2.4 and 2.5, we obtain the desired extension of Theorem 1.3.

Corollary 2.6 *Let Ω be an open set of $\mathbf{R}$ with finite length, such that $\Gamma = \partial\Omega$ is h-Minkowski measurable for some $h \in G_d$. Then we have*

$$N(\lambda) = \varphi(\lambda) - c_{1,d}M(h;\Gamma)f(\sqrt{\lambda}) + o(f(\sqrt{\lambda})), \quad \text{as } \lambda \to \infty, \qquad (2.8)$$

where the positive constant $c_{1,d}$ is given, as in (1.5), by

$$c_{1,d} = 2^{-(1-d)}\pi^{-d}(1-d)(-\zeta(d)).$$

(Note that $c_{1,d} > 0$ since $-\zeta(d) > 0$ for $d \in (0,1)$.)

We will also characterize the situation when we obtain sharp remainder estimates. This will provide an extension of Theorem 1.4.

Theorem 2.7 *Let Ω be an open subset of $\mathbf{R}$ with finite length. Let $(l_j)_{j=1}^{\infty}$ be the associated sequence, and let $h \in G_d$ for some $d \in (0,1)$. Then the following assertions are equivalent:*

(1) $l_j \asymp g(j)$, as $j \to \infty$;

(2) $0 < M_*(h;\Gamma) \le M^*(h;\Gamma) < \infty$;

(3) $\sum_{j=1}^{\infty}\{l_jx\} \asymp f(x)$, as $x \to \infty$;

(4) $\varphi(\lambda) - N(\lambda) \asymp f(\sqrt{\lambda})$, as $\lambda \to \infty$.

Here, $\{\gamma\} = \gamma - [\gamma]$ denotes the fractional part of the real number γ. Further, "$v(x) \asymp w(x)$ as $x \to a$" means that there exist positive constants c_1, c_2 such that $c_1v(x) \le w(x) \le c_2v(x)$, for all x in some neighborhood of a.

Remark 2.8 (*a*) The above results extend their counterpart in [LaPo2, §2, pp. 45-47] to general gauge functions in G_d. We can recover all the corresponding results in [LaPo1,2] by taking $h(x) = x^d$. (Note that in this case, $f(x) = h(x) = x^d$ and $g(x) = x^{-1/d}$.)

(*b*) Since by assumption (H1), $\lim_{x \to 0^+} x/h(x) = 0$, it follows easily from (2.3) that $f(x) = o(x)$, as $x \to \infty$. Hence, for example, in (2.8), the term proportional to $f(\sqrt{\lambda})$ is really an asymptotic second term as $\lambda \to \infty$.

(*c*) We will establish an analogue of Theorem 2.7 for one-sided (rather than two-sided) estimates; see Theorem 3.13.

We also obtain the following more general version of Theorem 1.6 above, thereby partially extending one of the main results in [LaMa1,2]. (Compare with [LaMa2, Theorem 2.4].)

Theorem 2.9 *Suppose $h \in G_d$ is a differentiable function with $xh'(x)/h(x) \geq \mu > 0$, for all $x > 0$ and some constant μ. Furthermore, let $\rho = d + i\nu$ ($0 < d < 1, \nu \in \mathbf{R}$) be a zero of the Riemann zeta-function. Then we can construct an open set $\Omega \subset \mathbf{R}$ with finite length and with boundary $\Gamma = \partial\Omega$ which is not h-Minkowski measurable, but such that*

$$N(\lambda) = \varphi(\lambda) + cf(\sqrt{\lambda}) + o(f(\sqrt{\lambda}), \quad as \ \lambda \to \infty,$$

for some nonzero constant c. Moreover, $0 < M_*(h; \Gamma) < M^*(h; \Gamma) < \infty$.

It follows that the converse of Theorem 2.4 (or Corollary 2.6) cannot be true for this value of d. This is the case, in particular, in the 'midfractal case' when $d = 1/2$ since $\zeta(s)$ is known to have a zero (even infinitely many zeros) on the 'critical line' *Re* $s = 1/2$ (see, e.g., [Ti]).

Remark 2.10 We can check that the functions in Example 2.3 satisfy the additional condition $xh'(x)/h(x) \geq \mu > 0$ for some positive constant μ. (See Examples 1 and 2 in the appendix.)

2.2   Higher dimensional case

Let n be any integer ≥ 1.

We will make the following assumption on the gauge function h:

(C1). $h : (0, \infty) \to (0, \infty)$ is a positive nondecreasing function. Moreover, $\lim_{x \to 0+} h(x) = 0, \lim_{x \to \infty} h(x) = \infty$.

(C2). There exist constants k_1, k_2, with $2^{n-1} < k_1 \leq k_2 \leq 2^n$, such that

$$k_1 h(x) \leq h(2x) \leq k_2 h(x), \quad \text{for all } x \text{ small.}$$

(C3). $h(x) \leq cx^{n-1}$ for some constant $c > 0$ and for all x small.

(C4). $h(x)/x^n \to 0$ as $x \to \infty$.

It is easy to check that when $n = 1$, hypothesis (H1)—(H3) made in Section 2.1 implies (C1)—(C3).

Example 2.11 There are many functions satifying (C1)—(C4). For instance, we can take

$$h(x) = \frac{x^d}{(\ln(\frac{1}{x}+1))^a} \quad \text{or} \quad h(x) = \frac{x^d}{\left(\ln(\ln(\frac{1}{x}+1)+1)\right)^a}$$

for some $d \in (n-1, n)$ and $a \geq 0$. The proof is included in Examples 3 and 4 of the appendix, where more general gauge functions involving iterated logarithms can also be found.

We can now state our extension of Theorem 1.2 to this setting.

Theorem 2.12 (error estimate) *Suppose that h satisfies* (C1)—(C4). *Let $\Omega \subset \mathbf{R}^n$ be an open set with finite volume and with boundary $\Gamma = \partial\Omega$ of finite upper h-Minkowski content; i.e., $M^*(h; \Gamma) < \infty$. Then we have*

$$N(\lambda) = \varphi(\lambda) + O(f(\sqrt{\lambda})), \quad as \ \lambda \to \infty, \tag{2.9}$$

where $\varphi(\lambda) = (2\pi)^{-n}\mathcal{B}_n|\Omega|_n\lambda^{n/2}$ and $f(x) = 1/h(1/x)$.

We will see in Example 7.7 that the above error estimate in (2.9) is *sharp* in general. Further, we will see, for instance in Examples 7.5 and 7.7, that there are many natural situations when $M^*(D; \Gamma) = \infty$ (or $M_*(D; \Gamma) = 0$), where D is the Minkowski dimension of $\Gamma = \partial\Omega$, while $0 < M_*(h; \Gamma) \leq M^*(h; \Gamma) < \infty$ for some suitable gauge function h in our class; so that our present estimates (in Section 2.1 or 2.2) genuinely improve upon the results of [La1] (or [LaPo2]).

Remark 2.13 (*a*) Theorem 2.12 extends the 'fractal case' of [La1, Theorem 2.1, p. 479] for the Dirichlet Laplacian. (Of course, the latter result is recovered by letting $h(x) = x^d$ and so $f(x) = x^d$.) Together with Definitions 2.1 and 2.2 above, this provides a concrete realization of the suggestion made in [La1, Remark 2.4(e), p. 481] to extend the error estimate of [La1] to more general gauge functions.

(b) After this work was completed, we have learned that a result analogous to Theorem 2.12—also generalizing [La1, Theorem 2.1] along the lines of [La1, Remark 2.4(e)], but with gauge functions of the type of Example 2.11— was obtained independently by A. M. Caetano [Ce2, §5].

(c) Under suitable hypotheses on Ω, Theorem 2.12 admits a counterpart for Neumann boundary conditions, much as in [La1, Theorem 4.1, pp. 510-511]; see Remark 6.11(b) below. (In this case, $H_0^1(\Omega)$ must be replaced by the Sobolev space $H^1(\Omega)$ in the variational formulation of the corresponding boundary value problem.)

(d) For the sake of clarity, we will work in this paper with gauge functions h defined on all of $(0, \infty)$. It is obvious, however, that we could also work with (suitable) h defined on (b, ∞), for some $b > 0$. This would enable us, in particular, to simplify the expression of some of the gauge functions studied in our examples.

(e) The expression '$\sqrt{\lambda}$', which appears in (2.9) as well as throughout Section 2.1, is quite natural from a physical point of view. Indeed, it can be thought of as the *frequency* of the 'fractal drum', associated with the eigenvalue λ. Accordingly, '$1/\sqrt{\lambda}$', which appears for instance on the right side of (2.9) in the expression $f(\sqrt{\lambda}) = 1/h(1/\sqrt{\lambda})$, represents the *wavelength* of the corresponding vibration.

The proofs of these theorems (in Sections 2.1 and 2.2) will often follow the ones provided in [La1], [LaMa2] and [LaPo2]. However, since the gauge functions h and the corresponding functions g and f are not explicitly known, we will see that in many cases, especially in Sections 3–5, the proofs are considerably more involved.

The main results of this paper are announced in [HeLa].

The rest of the paper is organized as follows:

We begin by considering the one-dimensional case (i.e., $n = 1$), as discussed in Section 2.1 above. This is done in Sections 3–5 which constitute the heart of this work.

More specifically, in Section 3, we prove, in particular, Theorem 2.7, which yields a two-sided error estimate and its converse for $N(\lambda)$ when $n = 1$ (see Section 3.2). We then obtain analogous results for one-sided estimates (see Section 3.3).

In Section 4, we establish Theorem 2.5, which provides a characterization of h-Minkowski measurability of $\Gamma = \partial\Omega$, with $\Omega \subset \mathbf{R}$ (see Section 4.1). we also prove Theorem 2.4 (and hence Corollary 2.6) showing the existence of a monotonic asymptotic second term for $N(\lambda)$; see Section 4.2. In the process, we establish (as in [LaPo1,2]) connections with the Riemann zeta-function $\zeta = \zeta(s)$ in the 'critical interval' $0 < Re\ s = d < 1$.

In Section 5, we prove Theorem 2.9, which yields a counterexample to the converse of Theorem 2.4 (provided that $\zeta(d + i\nu) = 0$ for some $\nu \in \mathbf{R}$).

Then, in Section 6, we consider the higher dimensional case discussed in Section 2.2; we thereby establish Theorem 2.12, which provides (sharp) remainder estimates for $N(\lambda)$ when $n \geq 2$.

Moreover, in Section 7, we present several examples illustrating our results; this shows in particular that some of our assumptions are necessary and that our estimates are in general optimal.

Finally, in the appendix, we provide detailed calculations showing that there is a large family of gauge functions which satisfy our hypotheses, both when $n = 1$ and $n \geq 2$.

3 Sharp Error Estimates and their Converse when $n = 1$

In the present section, as well as in Sections 4 and 5, we work as in Section 2.1 in the one-dimensional case (i.e., when $n = 1$). After establishing some preliminary results in Section 3.1, we will obtain a characterization of two-sided estimates in Section 3.2 (thereby proving Theorem 2.7), and of one-sided estimates in Section 3.3 (see Theorem 3.13).

3.1 Preliminaries

Before proceeding with the proof of the theorems, we will establish the following lemmas which will be used throughout the paper.

Recall that for $h \in G_d$, we set $g(x) = h^{-1}(1/x)$ and $f(x) = 1/h(1/x)$.

Lemma 3.1 *Let $h \in G_d$ for some $d \in (0, 1)$. Then we have:*

(1)

$$\lim_{x \to \infty} \frac{f(tx)}{f(x)} = t^d,$$

uniformly in t on any compact subset of $(0, \infty)$.

Futher, there exist constants $M_1 > 0$, $x_1 > 0$, $t_1 \geq 1$ such that

$$\frac{f(tx)}{f(x)} \leq M_1 t^\tau \quad \text{for all } x > x_1, \ t > t_1. \tag{3.1}$$

(2)

$$\lim_{x \to \infty} \frac{g(tx)}{g(x)} = t^{-1/d}, \tag{3.2}$$

uniformly in t on any compact subset of $(0, \infty)$.

Moreover, there exist constants $M_2 > 0$, $x_2 > 0$, $t_2 \geq 1$ such that

$$\frac{g(tx)}{g(x)} \leq M_2 t^{-1/\tau} \quad \text{for all } \ x > x_2, \ t > t_2. \tag{3.3}$$

14

Proof. (1) We easily check that

$$
\lim_{x\to\infty} \frac{f(tx)}{f(x)} = \lim_{x\to\infty} \frac{\frac{1}{h(1/tx)}}{\frac{1}{h(1/x)}} = \lim_{x\to\infty} \frac{h(1/x)}{h(1/tx)}
$$

$$
= \lim_{y\to 0^+} \frac{h(y)}{h(y/t)} \quad (\text{where } y = 1/x)
$$

$$
= t^d.
$$

The convergence is uniform in t on any compact subset of $(0,\infty)$ since if t is contained in some compact subset of $(0,\infty)$, then $1/t$ is also contained in the same type of set. Thus by assumption (H2), $h(y)/h(y/t)$ converges to t^d uniformly in t.

Next, we want to show that $f(tx)/f(x) \le M_1 t^\tau$ for some $M_1 > 0$ and all x, t large. By hypothesis (H3), we have for some $m > 0$ and $\tau \in (0,1)$,

$$
\frac{h(tx)}{h(x)} \ge m t^\tau, \quad \text{for all } 0 < x < x_0,\ 0 < t < t_0.
$$

We see that

$$
\frac{f(tx)}{f(x)} = \frac{\frac{1}{h(1/tx)}}{\frac{1}{h(1/x)}} = \frac{h(1/x)}{h(1/tx)} = \frac{h(y)}{h(sy)},
$$

where $y = 1/x$ and $s = 1/t$.

So if $0 < y < x_0$ (i.e., $x > 1/x_0$) and $0 < s < t_0$ (i.e., $t > 1/t_0$), we have

$$
\frac{h(y)}{h(sy)} = \frac{1}{\frac{h(sy)}{h(y)}} \le \frac{1}{m s^\tau} = \frac{1}{m} t^\tau.
$$

Hence we obtain that for $M_1 := 1/m$, $x > x_1 := 1/x_0$, $t > t_1 := \max(1, 1/t_0)$,

$$
\frac{f(tx)}{f(x)} \le M_1 t^\tau.
$$

(2) We will first prove that

$$
\lim_{x\to 0^+} \frac{h^{-1}(tx)}{h^{-1}(x)} = t^{1/d}, \tag{3.4}
$$

uniformly in t on any compact subset of $(0, \infty)$. Thus, we need to show that for any given $\epsilon_0 > 0$, there exists $x_0 > 0$ (depending only on ϵ_0) such that for all t in some compact subset of $(0, \infty)$, say $[a, b]$, and for all $x < x_0$, we have

$$t^{1/d}(1 - \epsilon_0)^{1/d} h^{-1}(x) < h^{-1}(tx) < t^{1/d}(1 + \epsilon_0)^{1/d} h^{-1}(x).$$

Let us consider the function $h(t^{1/d}(1+\epsilon_0)^{1/d} h^{-1}(x))$. For fixed $\epsilon_0 \in (0, 1)$, $t^{1/d}(1 + \epsilon_0)^{1/d}$ is also contained in some compact subset of $(0, \infty)$. Thus

$$\frac{h(t^{1/d}(1 + \epsilon_0)^{1/d} h^{-1}(x))}{h(h^{-1}(x))} = \frac{h(t^{1/d}(1 + \epsilon_0)^{1/d} h^{-1}(x))}{x}$$
$$\rightarrow (t^{1/d}(1 + \epsilon_0)^{1/d})^d = t(1 + \epsilon_0),$$

as $x \rightarrow 0^+$, uniformly in t on $[a, b]$. Hence if we take $\epsilon_1 = \frac{1}{2}\frac{\epsilon_0}{\epsilon_0+1}$, there exists x_1 (depending only on ϵ_1, and thus on ϵ_0) such that for all $h^{-1}(x) < x_1$ (i.e., $x < h(x_1)$), the following inequalities hold:

$$tx < t(1 + \epsilon_0)(1 - \epsilon_1)x < h(t^{1/d}(1 + \epsilon_0)^{1/d} h^{-1}(x)) < t(1 + \epsilon_0)(1 + \epsilon_1)x.$$

Note that we have chosen ϵ_1 in such a way that

$$(1 + \epsilon_0)(1 - \epsilon_1) = 1 + \frac{1}{2}\epsilon_0 > 1.$$

Consequently, we obtain for all $x < h(x_1)$:

$$tx < h(t^{1/d}(1 + \epsilon_0)^{1/d} h^{-1}(x)).$$

By assumption (H1), this implies that

$$h^{-1}(tx) < t^{1/d}(1 + \epsilon_0)^{1/d} h^{-1}(x).$$

Similarly, by considering $h(t^{1/d}(1 - \epsilon_0)^{1/d} h^{-1}(x))$, we can prove that there exists $x_2 > 0$ such that for all $x < h(x_2)$,

$$h(t^{1/d}(1 - \epsilon_0)^{1/d} h^{-1}(x)) < tx,$$

which is equivalent to

$$t^{1/d}(1 - \epsilon_0)^{1/d} h^{-1}(x) < h^{-1}(tx).$$

Now let $x_0 = \min(h(x_1), h(x_2))$. Thus, we showed that for all $x < x_0$,

$$t^{1/d}(1 - \epsilon_0)^{1/d} h^{-1}(x) < h^{-1}(tx) < t^{1/d}(1 + \epsilon_0)^{1/d} h^{-1}(x),$$

and the choice of x_0 is independent of t in the compact set $[a, b]$. Since ϵ_0 is arbitrary, we have proved (3.4), and shown that the convergence is uniform in t on any compact subset of $(0, \infty)$.

Then we can easily deduce from (3.4) that

$$\lim_{x \to \infty} \frac{g(tx)}{g(x)} = \lim_{x \to \infty} \frac{h^{-1}(1/tx)}{h^{-1}(1/x)} = \lim_{y \to 0^+} \frac{h^{-1}(\frac{1}{t}y)}{h^{-1}(y)} = \left(\frac{1}{t}\right)^{1/d} = t^{-1/d},$$

uniformly in t on any compact subset of $(0, \infty)$. This proves (3.2).

Next we want to show that (3.3) holds. By (H3), we know that

$$\frac{h(tx)}{h(x)} \geq mt^\tau, \quad \text{for all } 0 < x < x_0 \leq 1,\ 0 < t < t_0 \leq 1.$$

Now suppose $h^{-1}(y) < x_0$ (i.e., $y < h(x_0)$) and $s < mt_0^\tau$ (i.e., $(s/m)^{1/\tau} < t_0$). Then we have

$$\frac{h((\frac{s}{m})^{1/\tau} h^{-1}(y))}{h(h^{-1}(y))} = \frac{h((\frac{s}{m})^{1/\tau} h^{-1}(y))}{y} \geq m \left(\left(\frac{s}{m}\right)^{1/\tau}\right)^\tau = s;$$

that is,

$$h\left(\left(\frac{s}{m}\right)^{1/\tau} h^{-1}(y)\right) \geq sy, \quad \text{for all } y < h(x_0) \text{ and } s < mt_0^\tau.$$

Thus

$$\left(\frac{s}{m}\right)^{1/\tau} h^{-1}(y) \geq h^{-1}(sy).$$

Set $x = 1/y$, $t = 1/s$. Then for all $x > 1/h(x_0)$, $t > t_0^{-\tau}/m$, we have

$$\left(\frac{1}{tm}\right)^{1/\tau} h^{-1}\left(\frac{1}{x}\right) \geq h^{-1}\left(\frac{1}{tx}\right);$$

that is,

$$\left(\frac{1}{tm}\right)^{1/\tau} g(x) \geq g(tx).$$

So we obtain that

$$\frac{g(tx)}{g(x)} \leq M_2 t^{-1/\tau},$$

for all $x > 1/h(x_0) =: x_2$, $t > t_2 := \max(1, t_0^{-\tau}/m)$, and with $M_2 := m^{-1/\tau}$. This yields (3.3) and concludes the proof of the lemma. ∎

Proposition 3.2 *Suppose $h \in G_d$ for some $d \in (0,1)$, and let $g(x) = h^{-1}(1/x)$ and $f(x) = 1/h(1/x)$, as in (2.3). Then the following holds:*

(i) Suppose $\{a_j\}_{j=1}^{\infty}$ is a 'strongly bounded' sequence in $(0,\infty)$ (that is, there exist some positive constants c_1, c_2 such that $c_1 \le a_j \le c_2$, for all $j \ge 1$) and let $(b_j)_{j=1}^{\infty}$ in $(0,\infty)$ with $b_j \to 0$ as $j \to \infty$. Then

$$\lim_{j \to \infty} \frac{h(a_j b_j)}{a_j^d h(b_j)} = 1; \tag{3.5}$$

i.e., $h(a_j b_j) \sim a_j^d h(b_j)$ as $j \to \infty$.

In particular, if $a_j \to a \in (0,\infty)$ as $j \to \infty$, we have

$$\lim_{j \to \infty} \frac{h(a_j b_j)}{h(b_j)} = a^d. \tag{3.6}$$

(ii) If $(a_j)_{j=1}^{\infty}$ is a 'strongly bounded' sequence in $(0,\infty)$ and $\lim_{j \to \infty} b_j = \infty$, we have

$$\lim_{j \to \infty} \frac{g(a_j b_j)}{a_j^{-1/d} g(b_j)} = 1. \tag{3.7}$$

In particular, if $\lim_{j \to \infty} a_j = a \in (0,\infty)$, then

$$\lim_{j \to \infty} \frac{g(a_j b_j)}{g(b_j)} = a^{-1/d}. \tag{3.8}$$

(iii)

$$\frac{\int_{tx}^{\infty} g(u)du}{\int_{x}^{\infty} g(u)du} \to t^{1-1/d}, \quad as \ x \to \infty, \tag{3.9}$$

uniformly in t on any compact subset of $(0,\infty)$.

(iv) If $(a_j)_{j=1}^{\infty}$ is a 'strongly bounded' sequence in $(0,\infty)$ and $\lim_{j \to \infty} b_j = \infty$, we have

$$\frac{\int_{a_j b_j}^{\infty} g(u)du}{a_j^{1-1/d} \int_{b_j}^{\infty} g(u)du} \to 1, \quad as \ j \to \infty. \tag{3.10}$$

In particular, if $\lim_{j \to \infty} a_j = a \in (0,\infty)$, then

$$\frac{\int_{a_j b_j}^{\infty} g(u)du}{\int_{b_j}^{\infty} g(u)du} \to a^{1-1/d}, \quad as \ j \to \infty. \tag{3.11}$$

Proof. (i) follows directly from hypothesis (H2).

We know that for any given $\epsilon > 0$, there exists $x_0 > 0$ such that for all $t \in [c_1, c_2]$ and all $x < x_0$,

$$\left| \frac{h(tx)}{t^d h(x)} - 1 \right| < \epsilon.$$

Now since $\lim_{j \to \infty} b_j = 0$, there exists some $J_0 \geq 1$ such that for all $j > J_0$, $b_j < x_0$. Also, by our assumption that $a_j \in [c_1, c_2]$ for all $j \geq 1$, we see that for all $j > J_0$ (where J_0 only depends on x_0, which in turn only depends on ϵ),

$$\left| \frac{h(a_j b_j)}{a_j^d h(b_j)} - 1 \right| < \epsilon.$$

Thus we have proved (3.5), and hence also (3.6) as a special case.

(ii) can be proved in the same way as (i), in view of Lemma 3.1.

(iii) We know from Lemma 3.1(2) that for any given $\epsilon > 0$, there exists x_0 (only depending on ϵ) such that for all $x > x_0$ and for all t in a given compact subset of $(0, \infty)$, we have

$$(1 - \epsilon)t^{-1/d}g(x) < g(tx) < (1 + \epsilon)t^{-1/d}g(x).$$

So for any $x > x_0$, we see that

$$\frac{\int_{tx}^{\infty} g(u)du}{\int_x^{\infty} g(u)du} = \frac{t\int_x^{\infty} g(tv)dv}{\int_x^{\infty} g(u)du} \quad (\text{ where } v = u/t)$$

$$< t(1 + \epsilon)t^{-1/d}\frac{\int_x^{\infty} g(v)dv}{\int_x^{\infty} g(u)du} = (1 + \epsilon)t^{1-1/d}.$$

(The fact that the integrals involved are finite will be justified in the proof of Lemma 3.3(1) below.)

Similarly, we obtain that

$$\frac{\int_{tx}^{\infty} g(u)du}{\int_x^{\infty} g(u)du} > (1 - \epsilon)t^{1-1/d}, \quad \text{for all } x > x_0.$$

Thus we conclude that

$$\lim_{x \to \infty} \frac{\int_{tx}^{\infty} g(u)du}{\int_x^{\infty} g(u)du} = t^{1-1/d},$$

uniformly in t on any compact subset of $(0, \infty)$.

(iv) follows easily from (iii). It can be proved just like (i). ∎

Lemma 3.3 *Suppose* $h \in G_d$ *for some* $d \in (0, 1)$ *and* g, f *are as in* (2.3). *Then*

(1)
$$\lim_{x \to \infty} \frac{\int_x^\infty g(u) du}{x g(x)} = \frac{d}{1 - d}. \tag{3.12}$$

(2) *Furthermore,*

$$\lim_{x \to \infty} \frac{\int_x^\infty f(u) u^{-z-1} du}{x^{-z} f(x)} = \frac{1}{z - d}, \quad \text{for all } z \in \mathbf{C} \text{ with } \operatorname{Re} z > \tau, \tag{3.13}$$

where $\tau \in (0, 1)$ *is given as in hypothesis* (H3).

Proof. (1) With the notation of (3.3), we see that for any fixed $x > x_2$,

$$\begin{aligned}
\frac{\int_x^\infty g(u) du}{x g(x)} &= \frac{\int_1^\infty g(tx) x \, dt}{x g(x)} \quad \text{(where } t = u/x) \\
&= \int_1^\infty \frac{g(tx)}{g(x)} dt = \int_1^{t_2} \frac{g(tx)}{g(x)} dt + \int_{t_2}^\infty \frac{g(tx)}{g(x)} dt.
\end{aligned}$$

The fact that the integral $\int_x^\infty g(u) du$ is finite follows from the proof below.

For the first integral, since by Lemma 3.1(2) $g(tx)/g(x) \to t^{-1/d}$ uniformly on $[1, t_2]$ as $x \to \infty$, we have

$$\int_1^{t_2} \frac{g(tx)}{g(x)} dt \to \int_1^{t_2} t^{-1/d} dt = \frac{d}{1 - d} (1 - t_2^{1-1/d}), \quad \text{as } x \to \infty. \tag{3.14}$$

For the second integral, since (3.3) holds and

$$M_2 \int_{t_2}^\infty t^{-1/\tau} dt < \infty,$$

we can apply the Lebesgue Dominated Convergence Theorem to deduce from (3.2) that

$$\int_{t_2}^\infty \frac{g(tx)}{g(x)} dt \to \int_{t_2}^\infty t^{-1/d} dt = \frac{d}{1 - d} t_2^{1-1/d}, \quad \text{as } x \to \infty. \tag{3.15}$$

Adding (3.14) and (3.15), we obtain

$$\int_1^\infty \frac{g(tx)}{g(x)} dt \to \frac{d}{1-d}, \quad \text{as } x \to \infty;$$

which proves (3.12).

(2) We can prove (3.13) in a way very similar to (3.12). In view of (3.1), we have for any fixed $x > x_1$:

$$\frac{\int_x^\infty f(u)u^{-z-1}du}{x^{-z}f(x)} = \frac{\int_1^\infty f(tx)t^{-z-1}x^{-z-1}xdt}{x^{-z}f(x)} \quad \text{(where } t = u/x\text{)}$$

$$= \int_1^\infty \frac{f(tx)}{f(x)} t^{-z-1} dt$$

$$= \int_1^{t_2} \frac{f(tx)}{f(x)} t^{-z-1} dt + \int_{t_2}^\infty \frac{f(tx)}{f(x)} t^{-z-1} dt.$$

For the first integral, since by Lemma 3.1(1) $f(tx)/f(x) \to t^d$ as $x \to \infty$, uniformly on the interval $[1, t_1]$, we see that as $x \to \infty$,

$$\int_1^{t_1} \frac{f(tx)}{f(x)} t^{-z-1} dt \to \int_1^{t_1} t^{-z-1+d} dt = \frac{1}{d-z}(t_1^{-z+d} - 1). \qquad (3.16)$$

For the second integral, since by (3.1)

$$\left| \frac{f(tx)}{f(x)} t^{-z-1} \right| \leq M_1 |t^{-\mathrm{Re}\, z-1+\tau}|$$

and

$$\int_{t_1}^\infty t^{-\mathrm{Re}\, z-1+\tau} dt < \infty \quad \text{(since we assumed that } 0 < \tau < 1 \text{ and } \mathrm{Re}\, z > \tau\text{),}$$

Lemma 3.1(1) and the Lebesgue Dominated Convergence Theorem yield

$$\int_{t_1}^\infty \frac{f(tx)}{f(x)} t^{-z-1} dt \to \int_{t_1}^\infty t^{-z-1+d} dt = \frac{1}{d-z}(-t_1^{-z+d}), \quad \text{as } x \to \infty. \qquad (3.17)$$

(Note that it easily follows from (H2) and (H3) as well as from the above argument that $\tau \geq d$, so that $\mathrm{Re}\, z > d$, as required.)

Upon adding (3.16) and (3.17) together, we deduce that

$$\int_1^\infty \frac{f(tx)}{f(x)} t^{-z-1} dt \to \frac{1}{z-d}, \quad \text{as } x \to \infty,$$

which concludes the proof of (2). ∎

3.2   Two-sided estimates

We are now ready to characterize the situation when we obtain sharp estimates for $n = 1$. We first derive expressions for $M_*(h; \Gamma)$ and $M^*(h; \Gamma)$ that only depend on the sequence $(l_j)_{j=1}^\infty$ associated with Ω. Since $\Gamma_\epsilon \cap \Omega = \cup_{j=1}^\infty ((\partial I_j)_\epsilon \cap I_j)$ and $|(\partial I_j)_\epsilon \cap I_j| = \min(l_j, 2\epsilon)$, we have

$$|\Gamma_\epsilon \cap \Omega|_1 = \sum_{j=1}^\infty |(\partial I_j)_\epsilon \cap I_j|_1 = 2\epsilon J(2\epsilon) + \sum_{j > J(2\epsilon)} l_j, \qquad (3.18)$$

where

$$J(\epsilon) := \max\{j \geq 1 : l_j \geq \epsilon\}. \qquad (3.19)$$

Hence, if we let

$$L(h; \epsilon) := 2h(\epsilon) J(2\epsilon) + \epsilon^{-1} h(\epsilon) \sum_{j > J(2\epsilon)} l_j, \qquad (3.20)$$

it follows from Definition 2.1 that

$$M_*(h; \Gamma) = \liminf_{\epsilon \to 0^+} L(h; \epsilon); \qquad (3.21)$$

$$M^*(h; \Gamma) = \limsup_{\epsilon \to 0^+} L(h; \epsilon). \qquad (3.22)$$

In the rest of this paper, it will be convenient to adopt the following notation and terminology. If $h \in G_d$ and $(l_j)_{j=1}^\infty$ is any nonincreasing sequence of positive numbers tending to zero, we set

$$\begin{aligned} M_* &= M_*(h; (l_j)) := \liminf_{\epsilon \to 0^+} L(h; \epsilon); \\ M^* &= M^*(h; (l_j)) := \limsup_{\epsilon \to 0^+} L(h; \epsilon). \end{aligned} \qquad (3.23)$$

If $0 < M_* = M^* < \infty$, we write $M = M(h; (l_j))$ for this common value. In this case, we say that the *sequence* $(l_j)_{j=1}^\infty$ is *h-Minkowski measurable* and has *h-Minkowski content* M. Of course, if (l_j) is the sequence associated with the open set Ω, then $\Gamma = \partial\Omega$ is h-Minkowski measurable (in the sense of Definition 2.1) if and only if (l_j) is; in this case, the associated h-Minkowski contents are the same. Further, we set

$$\alpha := \liminf_{j \to \infty} \frac{l_j}{g(j)}, \quad \beta := \limsup_{j \to \infty} \frac{l_j}{g(j)}. \qquad (3.24)$$

We can now state the following result that will be used in the proofs of Theorems 2.7 and 2.4.

Theorem 3.4 *Given $h \in G_d$ for some $d \in (0,1)$, let $(l_j)_{j=1}^{\infty}$ be any positive nonincreasing sequence such that $0 < \alpha \leq \beta < \infty$ (i.e., $l_j \asymp g(j)$ as $j \to \infty$). Then*

(a)
$$\phi(\alpha, \beta) \leq M_* \leq M^* \leq \phi(\beta, \alpha), \tag{3.25}$$

where
$$\phi(u, v) := 2^{1-d}(u^d + \frac{d}{1-d}uv^{d-1}). \tag{3.26}$$

(b) *In particular, if $\alpha = \beta = L$, for some constant $L > 0$ (that is, $\lim_{j\to\infty} l_j/g(j) = L$), then $(l_j)_{j=1}^{\infty}$ is h-Minkowski measurable and its h-Minkowski content is given by*

$$M(h; (l_j)) = \frac{2^{1-d}}{1-d}L^d. \tag{3.27}$$

Proof. Clearly, part (b) follows from (a). So all we have to do is to establish part (a).

Let α_j be defined by $l_j = \alpha_j g(j) = \alpha_j h^{-1}(1/j)$. Thus, by (3.24), we see that

$$\alpha = \liminf_{\epsilon \to 0^+} \alpha_j, \quad \beta = \limsup_{\epsilon \to 0^+} \alpha_j. \tag{3.28}$$

From the definition (3.19) of $J(\epsilon)$, we have

$$J(2\epsilon) = \max\{j \geq 1 : l_j \geq 2\epsilon\} = \max\{j \geq 1 : \alpha_j g(j) \geq 2\epsilon\}$$
$$= \max\{j \geq 1 : h^{-1}(1/j) \geq \frac{2\epsilon}{\alpha_j}\} = \max\left\{j \geq 1 : j \leq \frac{1}{h\left(\frac{2\epsilon}{\alpha_j}\right)}\right\}.$$

We know that $\alpha(1 + o(1)) < \alpha_j < \beta(1 + o(1))$ as $j \to \infty$, and $h(x)$ is an increasing function, so

$$\frac{1}{h\left(\frac{2}{\alpha(1+o(1))}\epsilon\right)} \leq J(2\epsilon) \leq \frac{1}{h\left(\frac{2}{\beta(1+o(1))}\epsilon\right)}, \quad \text{as } \epsilon \to 0^+. \tag{3.29}$$

Note that since $g(x) = h^{-1}(1/x)$ is a positive and decreasing function,

$$\sum_{j>T} g(j) \sim \int_T^{\infty} g(x)dx, \quad \text{as } T \to \infty. \tag{3.30}$$

This is so, for example, because

$$\int_{T+1}^{\infty} g(x)dx < \sum_{j>T} g(j) < \int_{T}^{\infty} g(x)dx, \text{ for all } T > 0.$$

We also know that by Proposition 3.2(iii),

$$\frac{\int_{T+1}^{\infty} g(x)dx}{\int_{T}^{\infty} g(x)dx} \to 1, \text{ as } T \to \infty.$$

Thus (3.30) holds.

Next, in view of Proposition 3.2, we can proceed as follows:

$$
\begin{aligned}
\epsilon^{-1}h(\epsilon) \sum_{j>J(2\epsilon)} l_j &= \epsilon^{-1}h(\epsilon) \sum_{j>J(2\epsilon)} \alpha_j g(j) \\
&\leq \epsilon^{-1}h(\epsilon)\beta(1+o(1)) \sum_{j>J(2\epsilon)} g(j) \\
&\leq \epsilon^{-1}h(\epsilon)\beta(1+o(1)) \int_{J(2\epsilon)}^{\infty} g(x)dx \\
&\leq \epsilon^{-1}h(\epsilon)\beta(1+o(1)) \int_{h\left(\frac{1}{\frac{2}{\alpha(1+o(1))}\epsilon}\right)}^{\infty} g(x)dx \\
&\sim \epsilon^{-1}h(\epsilon)\beta(1+o(1)) \int_{\left(\frac{1}{\frac{2}{\alpha(1+o(1))}}\right)^d h(\epsilon)}^{\infty} g(x)dx \\
&\sim \epsilon^{-1}h(\epsilon)\beta(1+o(1)) \left(\frac{2}{\alpha(1+o(1))}\right)^{-d(1-1/d)} \int_{\frac{1}{h(\epsilon)}}^{\infty} g(x)dx \\
&= 2^{1-d}\alpha^{d-1}\beta(1+o(1))\epsilon^{-1}h(\epsilon) \int_{\frac{1}{h(\epsilon)}}^{\infty} g(x)dx, \qquad (3.31)
\end{aligned}
$$

where we used (3.9) in the step preceding the last equality. (Of course, the sign '$\sim$' holds as $\epsilon \to 0^+$. Further, the symbol '$o(1)$' denotes various functions vanishing as $\epsilon \to 0^+$.)

If we let $u = 1/h(\epsilon)$, then $\epsilon = h^{-1}(1/u) = g(u)$, and thus by (3.12),

$$\epsilon^{-1}h(\epsilon) \int_{\frac{1}{h(\epsilon)}}^{\infty} g(x)dx = \frac{\int_{u}^{\infty} g(x)dx}{ug(u)} \to \frac{d}{1-d}, \text{ as } u \to \infty \text{ (i.e., } \epsilon \to 0^+ \text{ by (H1))}.$$

Hence continuing with (3.31), we have

$$\epsilon^{-1} h(\epsilon) \sum_{j > J(2\epsilon)} l_j \leq 2^{1-d} \alpha^{d-1} \beta (1 + o(1)) \frac{d}{1 - d}, \quad \text{as } \epsilon \to 0^+. \tag{3.32}$$

Similarly, we can show that

$$\epsilon^{-1} h(\epsilon) \sum_{j > J(2\epsilon)} l_j \geq 2^{1-d} \beta^{d-1} \alpha (1 + o(1)) \frac{d}{1 - d}, \quad \text{as } \epsilon \to 0^+. \tag{3.33}$$

By (3.29), we see that

$$\frac{2h(\epsilon)}{h\left(\frac{2}{\alpha(1+o(1))}\epsilon\right)} \leq 2h(\epsilon) J(2\epsilon) \leq \frac{2h(\epsilon)}{h\left(\frac{2}{\beta(1+o(1))}\epsilon\right)}. \tag{3.34}$$

Thus, if we let $\epsilon \to 0^+$, then the left side of (3.34) is $\sim 2(\frac{\alpha(1+o(1))}{2})^d = 2^{1-d}\alpha^d(1 + o(1))$, and the right side is $\sim 2(\frac{\beta(1+o(1))}{2})^d = 2^{1-d}\beta^d(1 + o(1))$. So combining (3.20) and (3.32)—(3.34), we have as $\epsilon \to 0^+$,

$$(2^{1-d}\beta^{d-1}\alpha \frac{d}{1-d} + 2^{1-d}\alpha^d)(1 + o(1)) \leq L(h; \epsilon)$$

$$\leq (2^{1-d}\alpha^{d-1}\beta \frac{d}{1-d} + 2^{1-d}\beta^d)(1 + o(1)). \tag{3.35}$$

Now taking both the upper and lower limit in (3.35) as $\epsilon \to 0^+$, we obtain (3.25), in view of (3.23). ∎

From now on, for convenience, we introduce the following notation. If $(l_j)_{j=1}^{\infty}$ is any positive nonincreasing sequence such that $\sum_{j=1}^{\infty} l_j < \infty$, we let, for $x > 0$,

$$\delta(x) := (\sum_{j=1}^{\infty} l_j)x - \sum_{j=1}^{\infty} [l_j x] = \sum_{j=1}^{\infty} \{l_j x\}, \tag{3.36}$$

where, as before, $\{\gamma\} = \gamma - [\gamma]$ denotes the fractional part of the real number γ. Moreover, we let

$$\delta_* := \liminf_{x \to \infty} h(1/x)\delta(x) \quad \text{and} \quad \delta^* := \limsup_{x \to \infty} h(1/x)\delta(x). \tag{3.37}$$

We briefly comment on and motivate the definition of $\delta(x)$ in (3.36). With the notation of Section 2.1, we have in view of (2.2) and the fact that $\varphi(\lambda) := \pi^{-1}|\Omega|_1\lambda^{1/2} = (\sum_{j=1}^{\infty} l_j)x$ with $x := \sqrt{\lambda}/\pi$:

$$\delta(x) = \sum_{j=1}^{\infty}\{l_jx\} = \varphi(\lambda) - N(\lambda). \qquad (3.36')$$

Formula (3.36′), combined with the homogeneity properties of $h(x)$ and $f(x) = 1/h(1/x)$, will enable us to derive the asymptotic properties of $N(\lambda)$ (as $\lambda \to \infty$) from the corresponding ones for $\delta(x)$ (as $x \to \infty$). For this reason, we will now work with $\delta(x)$ instead of $N(\lambda)$.

Theorem 3.4 above contained the implication $(1) \Rightarrow (2)$ in Theorem 2.7. The next theorem contains the implication $(1) \Rightarrow (3)$.

Theorem 3.5 *Given $h \in G_d$ for some $d \in (0,1)$, let $(l_j)_{j=1}^{\infty}$ be an arbitrary positive nonincreasing sequence such that $0 < \alpha \leq \beta < \infty$. Then*

$$\frac{d}{1-d}\alpha\beta^{1-d} \leq \delta_* \leq \delta^* \leq \beta^d + \frac{d}{1-d}\beta\alpha^{1-d}. \qquad (3.38)$$

Proof. In view of (3.36) and (3.19), we have that

$$\begin{aligned}
\delta(x) &= \sum_{j\leq J(1/x)}\{l_jx\} + \sum_{j>J(1/x)}\{l_jx\} \\
&= \sum_{j\leq J(1/x)}\{l_jx\} + x\sum_{j>J(1/x)}l_j \\
&\leq J(1/x) + x\sum_{j>J(1/x)}l_j. \qquad (3.39)
\end{aligned}$$

(The second equality follows from the fact that for all $j > J(1/x)$, $l_j < 1/x$; that is, $l_jx < 1$, and so $\{l_jx\} = l_jx$.)

Since $0 \leq \{\gamma\} < 1$ for all γ, we have

$$\begin{aligned}
h\left(\frac{1}{x}\right)x\sum_{j>J(1/x)}l_j &< h\left(\frac{1}{x}\right)\delta(x) \\
&\leq h\left(\frac{1}{x}\right)J\left(\frac{1}{x}\right) + h\left(\frac{1}{x}\right)x\sum_{j>J(1/x)}l_j. \qquad (3.40)
\end{aligned}$$

If we let $2\epsilon = 1/x$, we deduce from (3.29) and hypothesis (H2) that

$$h\left(\frac{1}{x}\right) J\left(\frac{1}{x}\right) = h(2\epsilon)J(2\epsilon) \leq \frac{h(2\epsilon)}{h\left(\frac{2}{\beta(1+o(1))}\epsilon\right)}$$

$$\sim \beta^d(1 + o(1)), \text{ as } \epsilon \to 0^+. \tag{3.41}$$

Also, by (H2),

$$h\left(\frac{1}{x}\right) x \sum_{j>J(1/x)} l_j = \frac{1}{2}\epsilon^{-1}h(2\epsilon) \sum_{j>J(2\epsilon)} l_j$$

$$\leq \frac{1}{2}2^d(1 + o(1))\epsilon^{-1}h(\epsilon) \sum_{j>J(2\epsilon)} l_j$$

$$\leq 2^{d-1}\beta(1 + o(1))2^{1-d}\alpha^{d-1}\frac{d}{1-d}$$

$$= \beta(1 + o(1))\alpha^{d-1}\frac{d}{1-d}, \tag{3.42}$$

as $\epsilon \to 0^+$, where we used (3.32) in the last inequality. (Note that we may apply (3.32) from the proof of Theorem 3.4 because the assumptions of Theorems 3.4 and 3.5 are the same.)

Similarly, we can show (using now (3.33) instead of (3.32)) that

$$xh\left(\frac{1}{x}\right) \sum_{j>J(1/x)} l_j \geq \alpha(1 + o(1))\beta^{d-1}\frac{d}{1-d}, \tag{3.43}$$

as $\epsilon \to 0^+$ (i.e., as $x \to \infty$). Thus, in the light of (3.40)—(3.43), we have

$$\alpha(1 + o(1))\beta^{d-1}\frac{d}{1-d} \leq h\left(\frac{1}{x}\right)\delta(x)$$

$$\leq \beta^d(1 + o(1)) + \alpha^{d-1}\beta(1 + o(1))\frac{d}{1-d}, \tag{3.44}$$

as $x \to \infty$. Now taking both the upper and lower limit in (3.44), we deduce (3.38) from definition (3.37), as desired. ∎

The following result contains the implication (3) $\Rightarrow$ (1) in Theorem 2.7.

Theorem 3.6 *Let $l_1 \geq l_2 \geq \cdots > 0$ be such that $\sum_{j=1}^{\infty} l_j < \infty$ and let $\delta(x)$ be defined by (3.36). Assume that for some $h \in G_d$, $\delta(x) \asymp f(x)$ as $x \to \infty$. Then $l_j \asymp g(j)$ as $j \to \infty$. (Recall that $f(x) = 1/h(1/x)$ and $g(x) = h^{-1}(1/x)$.)*

Proof. By our assumption, there exist positive constants a_1, a_2 and x_1 such that for all $x > x_1$:

$$a_1 f(x) \le \delta(x) \le a_2 f(x). \tag{3.45}$$

Suppose ϵ_0 is a given small number, and let $k \ge 2$ be a fixed integer with

$$k > \left(\frac{2a_2(1 + \epsilon_0)}{a_1} \right)^{\frac{1}{1-d}}.$$

Then by Lemma 3.1(1), there exists $x_2 > 0$ such that for all $x > x_2$:

$$k^d(1 - \epsilon_0) < \frac{f(kx)}{f(x)} < k^d(1 + \epsilon_0) < \frac{\frac{1}{2} a_1 k}{a_2}. \tag{3.46}$$

Now set $x_0 = \max(x_1, x_2)$. Let

$$U := \sum_{\{l_j x\} < k^{-1}} \{l_j x\} \quad \text{and} \quad V := \sum_{\{l_j x\} \ge k^{-1}} \{l_j x\},$$

so that

$$\delta(x) = \sum_{j=1}^{\infty} \{l_j x\} = U + V. \tag{3.47}$$

Since $k\{\gamma\} = \{k\gamma\}$ for $\{\gamma\} < k^{-1}$, it follows from (3.45) and (3.46) that for $x \ge x_0$ (hence $kx > x_0$),

$$kU = \sum_{\{l_j x\} < k^{-1}} \{l_j kx\} \le \delta(kx)$$

$$\le a_2 f(kx) \le \frac{1}{2} a_1 k f(x) \le \frac{1}{2} k \delta(x).$$

Thus we have $U \le \frac{1}{2}\delta(x)$. So $V \ge \frac{1}{2}\delta(x)$. We then deduce from the last inequality that the number of integers j with $\{l_j x\} \ge k^{-1}$ is greater than $\frac{1}{2}\delta(x)$ (since each $\{l_j x\}$ is less than 1); since $\gamma \ge \{\gamma\}$, the same is true of the number of j with $l_j x \ge k^{-1}$ or, equivalently, $l_j \ge \frac{1}{kx}$.

Thus, with $J(\epsilon)$ defined as in (3.19), it follows from the above assertion that

$$J\left(\frac{1}{kx}\right) > \frac{1}{2}\delta(x) \ge \frac{1}{2} a_1 f(x), \quad \text{for all } x > x_0. \tag{3.48}$$

Since $l_j \downarrow 0$, there exists a positive integer j_1 such that $l_{j_1} < \frac{1}{kx_0}$ (k and x_0 are fixed for now); so that $x_0 < \frac{1}{kl_j}$ for all $j \geq j_1$. Fix $j \geq j_1$ and consider $x \in [x_0, \frac{1}{kl_j})$; thus $l_j < \frac{1}{kx}$. Then by definition of $J(\epsilon)$ and (3.48),

$$\frac{1}{2}a_1 f(x) < J\left(\frac{1}{kx}\right) < j. \tag{3.49}$$

This is true for all $x \in [x_0, \frac{1}{kl_j})$; so letting $x \to (\frac{1}{kl_j})^-$ in (3.49), we have

$$\frac{1}{2}a_1 f\left(\frac{1}{kl_j}\right) \leq j, \text{ which means } h(kl_j) \geq \frac{a_1}{2j}.$$

[This follows from the continuity (or the monotonicity) of h and hence of f, assumed in (H1)]. By (H1) and (3.2), this implies that

$$kl_j \;\geq\; h^{-1}\left(\frac{a_1}{2j}\right) = g\left(\frac{2}{a_1}j\right)$$

$$\geq\; \left(\frac{2}{a_1}\right)^{-1/d}(1+o(1))g(j), \text{ as } j \to \infty.$$

Therefore

$$\frac{l_j}{g(j)} = \alpha_j \geq \left(\frac{2}{a_1}\right)^{-1/d}\frac{1}{k}(1+o(1)), \text{ as } j \to \infty. \tag{3.50}$$

By taking the lower limit in (3.50) as $j \to \infty$, we obtain

$$\alpha = \liminf_{j \to \infty} \alpha_j \geq \left(\frac{2}{a_1}\right)^{-1/d}\frac{1}{k} > 0. \tag{3.51}$$

Next, let σ denote the number of $j \in (J(\frac{2}{x}), J(\frac{1}{x})]$ with $\{l_j x\} \geq 1/2$ and let κ denote the number of j in this interval with $\{l_j x\} < \frac{1}{2}$. Then $\sigma + \kappa = J(1/x) - J(2/x)$. By the definition of σ and $\delta(x)$, $\frac{1}{2}\sigma \leq \delta(x)$. Therefore, if j is counted by κ, $\{l_j \frac{1}{2}x\} \geq \frac{1}{2}$. (Indeed, $J(1/x) \geq j > J(2/x)$ implies that $\frac{1}{2} \leq \frac{1}{2}l_j x < 1$ and so $\{l_j \frac{1}{2}x\} = \frac{1}{2}l_j x \geq \frac{1}{2}$.) So $\frac{1}{2}\kappa \leq \delta(\frac{1}{2}x)$. Thus,

$$J\left(\frac{1}{x}\right) - J\left(\frac{2}{x}\right) \;=\; \sigma + \kappa \leq 2\delta(x) + 2\delta(\frac{1}{2}x)$$

$$\leq\; 2a_2 f(x) + 2a_2 f(\frac{1}{2}x). \tag{3.52}$$

Since we know that $\lim_{x \to \infty} f(x/2)/f(x) = 2^{-d}$ and $\frac{1}{2} < 2^{-d} < 1$, there exist constants c_1, c_2 with $\frac{1}{2} < c_1 < c_2 < 1$ and $x_3 > 0$ such that for all $x \geq x_3$:

$$c_1 f(x) \leq f\left(\frac{x}{2}\right) \leq c_2 f(x).$$

Let the previous x_0 be replaced by $\max\{x_1, x_2, x_3\}$, but still be called x_0. Continuing with (3.52), we have for all $x \geq 2x_0$:

$$J\left(\frac{1}{x}\right) - J\left(\frac{2}{x}\right) \leq 2a_2(1 + c_2)f(x). \tag{3.53}$$

For any fixed $x \geq 2x_0$, let $m = m(x)$ be the integer such that $\frac{x}{2^m} \geq x_0 > \frac{x}{2^{m+1}}$. Since $\frac{x}{2^{m-1}} \geq 2x_0$, we can apply the estimate (3.53) with x replaced by $\frac{x}{2^k}$, for $k = 0, 1, \ldots, m-1$, and since $\frac{1}{2x_0} < \frac{2^m}{x}$, we have

$$\begin{aligned}
J\left(\frac{1}{x}\right) &= \sum_{k=0}^{m-1}\left(J\left(\frac{2^k}{x}\right) - J\left(\frac{2^{k+1}}{x}\right)\right) + J\left(\frac{2^m}{x}\right) \\
&\leq \sum_{k=0}^{m-1} 2a_2(1+c_2)f\left(\frac{x}{2^k}\right) + J\left(\frac{1}{2x_0}\right) \\
&\leq 2a_2(1+c_2)\sum_{k=0}^{m-1} c_2^k f(x) + J\left(\frac{1}{2x_0}\right) \\
&\leq 2a_2(1+c_2)\frac{1}{1-c_2}f(x) + J\left(\frac{1}{2x_0}\right) \quad \text{(since } c_2 < 1\text{).} \tag{3.54}
\end{aligned}$$

Let $J_0 := J(\frac{1}{2x_0})$, and let $j > J_0$, $x = 1/l_j$, so that $x > 2x_0$. Then we deduce from (3.54) and (3.19) that

$$J_0 < j \leq J\left(\frac{1}{x}\right) \leq 2a_2\frac{1+c_2}{1-c_2}f(x) + J_0 = cf(x) + J_0,$$

where c is the positive constant $2a_2(1+c_2)/(1-c_2)$. Thus we obtain

$$0 < j - J_0 \leq cf(x) = cf\left(\frac{1}{l_j}\right) = c\frac{1}{h(l_j)};$$

that is, $h(l_j) \leq c/(j - J_0)$. By assumption (H1), this means that

$$l_j \leq h^{-1}\left(c\frac{1}{j-J_0}\right) = g\left(\frac{1 - J_0/j}{c}j\right). \tag{3.55}$$

This is true for all $j > J_0$. Hence we see that

$$\beta = \limsup_{j \to \infty} \frac{l_j}{g(j)} \leq \limsup_{j \to \infty} \frac{g\left(\frac{1 - J_0/j}{c}j\right)}{g(j)} = c^{1/d} < \infty.$$

Note that the last equality follows from (3.8). Together with (3.51), this proves the theorem. ∎

Remark 3.7 We used both $\delta_* > 0$ and $\delta^* < \infty$ to prove $\alpha > 0$. In fact, we can find some examples in [LaPo2, §3.3] showing that it is possible that $\delta_* > 0$ but $\alpha = 0$. On the other hand, the assumption $\delta^* < \infty$ is sufficient to guarantee $\beta < \infty$.

The next result contains part of the implication $(2) \Rightarrow (1)$ in Theorem 2.7.

Theorem 3.8 *Suppose that $l_1 \geq l_2 \cdots > 0$ and $0 < M_* \leq M^* < \infty$, for some $h \in G_d$. Then $\alpha > 0$.*

In order to prove this theorem, we shall need the following lemma.

Lemma 3.9 *If $\liminf_{\epsilon \to 0^+} h(\epsilon)J(\epsilon) > 0$,   then $\alpha > 0$.*

Proof. Let $\alpha_j = l_j/g(j)$ as before. If $l_j = l_{j+1}$, then $\alpha_j < \alpha_{j+1}$; thus

$$\alpha = \liminf_{j \to +\infty} \alpha_j = \liminf_{j: l_j > l_{j+1}} \alpha_{j+1}. \tag{3.56}$$

Suppose $\liminf_{\epsilon \to 0^+} h(\epsilon)J(\epsilon) > 0$. Then there is some constant $C > 0$ with $h(\epsilon)J(\epsilon) \geq C$ for all $\epsilon > 0$. Say $l_j > l_{j+1}$. Then for any $\epsilon \in (l_{j+1}, l_j)$, we have $J(\epsilon) = j$ and hence $h(\epsilon)j \geq C$. That is,

$$\epsilon \geq h^{-1}(C/j) = g(j/C).$$

Since this is true for all $\epsilon \in (l_{j+1}, l_j)$, we may let $\epsilon \to l_{j+1}^+$ and obtain

$$l_{j+1} \geq g(j/C) > g(\frac{j+1}{C}),$$

because g is a decreasing function. Then we deduce that

$$
\begin{aligned}
\alpha &= \liminf_{\epsilon \to 0^+} \alpha_j = \liminf_{j:\, l_j > l_{j+1}} \alpha_{j+1} = \liminf_{j:\, l_j > l_{j+1}} \frac{l_{j+1}}{g(j+1)} \\
&\geq \liminf_{j:\, l_j > l_{j+1}} \frac{g(\frac{1}{C}(j+1))}{g(j+1)} = \left(\frac{1}{C}\right)^{-1/d} = C^{1/d} > 0, \qquad (3.57)
\end{aligned}
$$

by (3.2). This proves the lemma. ∎

Proof of Theorem 3.8. Since $M^* < \infty$, it follows from definitions (3.20) and (3.23) that there is some finite constant T such that

$$
\epsilon^{-1} h(\epsilon) \sum_{j > J(2\epsilon)} l_j \leq T, \quad \text{for all } \epsilon > 0. \qquad (3.58)
$$

From Lemma 3.9, in order to prove $\alpha > 0$, it is sufficient to show that $\liminf_{\epsilon \to 0^+} h(\epsilon) J(\epsilon) > 0$. Suppose not. Then there exists a decreasing sequence $(\epsilon_q)_{q=1}^{\infty}$ with $\epsilon_q \to 0$ as $q \to \infty$ such that $\lim_{q \to \infty} h(\epsilon_q) J(\epsilon_q) = 0$. Let

$$
\theta_q := h(\epsilon_q) J(\epsilon_q). \qquad (3.59)
$$

So $\lim_{q \to \infty} \theta_q = 0$. Therefore we may assume that $\theta_q < 1$ for all $q \geq 1$. Let

$$
\eta_q := (\theta_q)^{-1/2} \epsilon_q. \qquad (3.60)
$$

Thus $2\eta_q > 2\epsilon_q$ and $J(2\eta_q) \leq J(2\epsilon_q) \leq J(\epsilon_q)$. We then have, by (3.20),

$$
\begin{aligned}
L(h; \eta_q) &= \eta_q^{-1} h(\eta_q) \sum_{j > J(2\eta_q)} l_j + 2h(\eta_q) J(2\eta_q) \\
&= \eta_q^{-1} h(\eta_q) \sum_{J(2\eta_q) < j \leq J(2\epsilon_q)} l_j + \eta_q^{-1} h(\eta_q) \sum_{j > J(2\epsilon_q)} l_j + 2h(\eta_q) J(2\eta_q) \\
&=: E_q + F_q + G_q, \quad \text{say.} \qquad (3.61)
\end{aligned}
$$

First, we note that

$$
G_q := 2h(\eta_q) J(2\eta_q) \leq 2h(\eta_q) J(\epsilon_q). \qquad (3.62)
$$

By hypothesis (H3), we know that there exist some $\tau \in (0,1), m > 0$ such that for all x, t small enough,

$$
\frac{h(x)}{h(tx)} \leq \frac{1}{m} t^{-\tau}.
$$

Now since $\eta_q \to 0$ and $\theta_q \to 0$ as $q \to \infty$, there exists $q_0 \geq 1$ such that for all $q > q_0$, the following holds:

$$\frac{h(\eta_q)}{h(\epsilon_q)} = \frac{h(\eta_q)}{h(\theta_q^{1/2}\eta_q)} \leq \frac{1}{m}\theta_q^{-\frac{\tau}{2}};$$

that is,

$$h(\eta_q) \leq \frac{1}{m}\theta_q^{-\frac{\tau}{2}}h(\epsilon_q). \tag{3.63}$$

(To see that $\eta_q \to 0$, we may argue as follows: by (3.59) and (3.60),

$$\eta_q^2 = \frac{\epsilon_q^2}{\theta_q} = \frac{\epsilon_q}{h(\epsilon_q)} \cdot \epsilon_q \cdot \frac{1}{J(\epsilon_q)} \to 0$$

since $\epsilon_q \to 0$, $J(\epsilon_q) \to \infty$ and $\epsilon_q/h(\epsilon_q) \to 0$, as $q \to \infty$; the latter limit holds because hypothesis (H1) implies that $x/h(x) \to 0$ as $x \to 0^+$.)

Hence, continuing with (3.62), we deduce that for $q > q_0$:

$$\begin{aligned} G_q &\leq 2h(\eta_q)J(\epsilon_q) \leq 2\frac{1}{m}\theta_q^{-\frac{\tau}{2}}h(\epsilon_q)J(\epsilon_q) \\ &= \frac{2}{m}\theta_q^{1-\frac{\tau}{2}} < \frac{2}{m}\theta_q^{\frac{1-\tau}{2}}, \end{aligned} \tag{3.64}$$

since $\theta_q < 1$.

Similarly, since $l_j < 2\eta_q$ whenever $j > J(2\eta_q)$, we have

$$\begin{aligned} E_q &:= \eta_q^{-1}h(\eta_q) \sum_{J(2\eta_q)<j\leq J(2\epsilon_q)} l_j \leq \eta_q^{-1}h(\eta_q)2\eta_q J(2\epsilon_q) \\ &\leq 2h(\eta_q)J(\epsilon_q) < \frac{2}{m}\theta_q^{\frac{1-\tau}{2}}, \end{aligned} \tag{3.65}$$

as in (3.64).

Finally, we deduce from (3.63) that for $q > q_0$:

$$\begin{aligned} F_q &:= \eta_q^{-1}h(\eta_q) \sum_{j>J(2\epsilon_q)} l_j = \theta_q^{1/2}\epsilon_q^{-1}h(\eta_q) \sum_{j>J(2\epsilon_q)} l_j \\ &\leq \theta_q^{1/2}\frac{1}{m}\theta_q^{-\frac{\tau}{2}}h(\epsilon_q)\epsilon_q^{-1} \sum_{j>J(2\epsilon_q)} l_j \leq \frac{1}{m}T\theta_q^{\frac{1-\tau}{2}}, \end{aligned} \tag{3.66}$$

where we have used (3.63) in the first and (3.58) in the last inequality.

Putting (3.64)—(3.66) into (3.61), we obtain for $q > q_0$:

$$0 < L(h; \eta_q) \leq \frac{1}{m}(T+4)\theta_q^{\frac{1-\tau}{2}}.$$

Since $\tau < 1$, $1 - \tau > 0$. We also know that $\lim_{q \to \infty} \theta_q = 0$. Thus it is clear that $L(h; \eta_q) \to 0$ as $q \to \infty$, which contradicts the assumption that $M_* > 0$.

Therefore, we must have $\liminf_{\epsilon \to 0^+} h(\epsilon) J(\epsilon) > 0$, which implies $\alpha > 0$ by Lemma 3.9. ∎

The following result is really a corollary of Remark 3.7 and of the proof of Theorem 3.8. It completes our proof of Theorem 2.7.

Theorem 3.10 *Let $l_1 \geq l_2 \geq \cdots > 0$ with $\sum_{j=1}^{\infty} l_j < \infty$. If $M^* < \infty$, then $\beta < \infty$.*

Proof. Recalling the definition (3.20) of $L(h; \epsilon)$, we deduce from inequality (3.40) that

$$
\begin{aligned}
h\left(\frac{1}{x}\right)\delta(x) &\leq h\left(\frac{1}{x}\right)J\left(\frac{1}{x}\right) + xh(\frac{1}{x})\sum_{j>J(1/x)} l_j \\
&= h(2\epsilon)J(2\epsilon) + \frac{1}{2}\epsilon^{-1}h(2\epsilon)\sum_{j>J(2\epsilon)} l_j \quad (\text{where } 2\epsilon = \frac{1}{x}) \\
&= 2^d(1+o(1))h(\epsilon)J(2\epsilon) + \frac{1}{2}\epsilon^{-1}2^d(1+o(1))h(\epsilon)\sum_{j>J(2\epsilon)} l_j \\
&= 2^{d-1}(1+o(1))(2h(\epsilon)J(2\epsilon) + \epsilon^{-1}h(\epsilon)\sum_{j>J(2\epsilon)} l_j) \\
&= 2^{d-1}(1+o(1))L(h; \epsilon), \quad\quad\quad (3.67)
\end{aligned}
$$

where $o(1) \to 0$ as $\epsilon \to 0^+$ (i.e., as $x \to \infty$).

Hence

$$
\begin{aligned}
\delta^* &= \limsup_{x \to \infty} h(1/x)\delta(x) \\
&\leq \limsup_{\epsilon \to 0^+} 2^{d-1}(1+o(1))L(h; \epsilon) = 2^{d-1}M^* < \infty.
\end{aligned}
$$

The fact that $\beta < \infty$ now follows from Remark 3.7. ∎

Remark 3.11 From the above proof, we deduce that $M^* < \infty$ implies $\delta^* < \infty$. Hence, under our present assumptions (H1)—(H3) on h, this yields the one-dimensional case of Theorem 2.12.

We close Section 3.2 by completing the proof of Theorem 2.7.

Proof of Theorem 2.7. Given the above results, all that remains to show is the equivalence of assertions (3) and (4) in Theorem 2.7. This is immediate, however, since by formula (3.36$'$) and Lemma 3.1(1), we have with $x := \sqrt{\lambda}/\pi$:

$$\varphi(\lambda) - N(\lambda) = \delta(x) \asymp f(x) = f\left(\frac{\sqrt{\lambda}}{\pi}\right)$$
$$= \pi^{-d} f(\sqrt{\lambda})(1 + o(1)) \asymp f(\sqrt{\lambda}),$$

as $x \to \infty$ (i.e., $\lambda \to \infty$). ∎

3.3   One-sided estimates

In the following, we will establish some further related results. We will obtain, in particular, an analogue of Theorem 2.7 for one-sided estimates. To do so, we shall need the following converse to Theorem 3.10.

Theorem 3.12 *Let $h \in G_d$ for some $d \in (0,1)$. Assume that $l_1 \geq l_2 \geq \cdots > 0$ and $\sum_{j=1}^{\infty} l_j < \infty$. Then $\beta < \infty$ implies $M^* < \infty$.*

Proof. Assume that $\beta < \infty$. Then, in view of definitions (3.20) and (3.23), we have

$$M^* = \limsup_{\epsilon \to 0^+}(2h(\epsilon)J(2\epsilon) + \epsilon^{-1}h(\epsilon) \sum_{j>J(2\epsilon)} l_j). \tag{3.68}$$

It follows from (3.34) that

$$h(\epsilon)J(\epsilon) \leq \frac{h(\epsilon)}{h\left(\frac{1}{\beta(1+o(1))}\epsilon\right)} \sim \beta^d(1 + o(1)), \quad \text{as } \epsilon \to 0^+,$$

by assumption (H2).

So there exists a constant $Q > 0$ such that

$$J(\epsilon) \leq \frac{Q}{h(\epsilon)}, \quad \text{for all } \epsilon > 0.$$

Since $J(2\epsilon) \leq J(\epsilon) \leq J(\epsilon/2) \leq \dots$, and $l_j < 2^{-(k-1)}\epsilon$ for all $j > J(2^{-(k-1)}\epsilon)$, we see that for all positive ϵ small enough,

$$
\begin{aligned}
\sum_{j > J(2\epsilon)} l_j &= \sum_{k=0}^{\infty} \sum_{j=J(2^{-k+1}\epsilon)}^{J(2^{-k}\epsilon)} l_j < \sum_{k=0}^{\infty} J(2^{-k}\epsilon) 2^{-k+1}\epsilon \\
&\leq \sum_{k=0}^{\infty} \frac{Q}{h(2^{-k}\epsilon)} 2^{-k+1}\epsilon = 2\epsilon Q \sum_{k=0}^{\infty} f(\frac{2^k}{\epsilon}) 2^{-k} \\
&\leq 2\epsilon Q f(\frac{1}{\epsilon}) \sum_{k=0}^{\infty} c_2^k 2^{-k} = 2\epsilon Q f(\frac{1}{\epsilon}) \frac{1}{1 - c_2/2} \\
&= \frac{2\epsilon Q}{h(\epsilon)} \frac{1}{1 - c_2/2}.
\end{aligned}
$$

(Recall that since $\lim_{x \to \infty} f(2x)/f(x) = 2^d$, we can find some constants c_1, c_2 with $1 < c_1 < c_2 < 2$ such that $c_1 f(x) \leq f(2x) \leq c_2 f(x)$ for all x large enough.) We stress that in the right-hand side of the first equality above, the summation over j is taken over all integers in the half-open, half-closed interval $(J(2^{-k+1}\epsilon), J(2^{-k}\epsilon)]$. When appropriate, we will use such a convention, most often implicitly, throughout the paper.

Thus we have

$$\epsilon^{-1} h(\epsilon) \sum_{j > J(2\epsilon)} l_j \leq \frac{2Q}{1 - c_2/2}, \quad \text{for all } \epsilon \text{ small enough,} \tag{3.69}$$

and

$$2h(\epsilon) J(2\epsilon) \leq \frac{2h(\epsilon)}{h(\frac{2}{\beta(1+o(1))}\epsilon)} \sim 2^{1-d} \beta^d (1 + o(1)), \quad \text{as } \epsilon \to 0^+. \tag{3.70}$$

So putting (3.69) and (3.70) together, we see that

$$M^* = \limsup_{\epsilon \to 0^+} \left(2h(\epsilon) J(2\epsilon) + \epsilon^{-1} h(\epsilon) \sum_{j > J(2\epsilon)} l_j \right) < \infty,$$

as desired. ∎

We can now state the counterpart of Theorem 2.7 for one-sided estimates, which extends [LaPo2, Theorem 3.10]. It shows in particular that when $n = 1$ and $h \in G_d$, the conclusion of Theorem 2.12 (that is, of the generalization of Theorem 1.2 [La1, Theorem 2.1]) is equivalent to its hypothesis.

Theorem 3.13 *Let $h \in G_d$, $g(x) = h^{-1}(1/x)$, $f(x) = 1/h(1/x)$ and let $(l_j)_{j=1}^{\infty}$ be an arbitrary positive nonincreasing sequence such that $\sum_{j=1}^{\infty} l_j < \infty$. Then the following statements are equivalent:*

(1) $l_j = O(g(j))$, as $j \to \infty$;

(2) $M^* < \infty$;

(3) $\delta(x) = O(f(x))$, as $x \to \infty$;

(4) $N(\lambda) = \varphi(\lambda) + O(f(\sqrt{\lambda}))$, as $\lambda \to \infty$, if $(l_j)_{j=1}^{\infty}$ is the sequence associated with the open set $\Omega \subset \mathbf{R}$, as in Section 2.1.

Proof. By Theorem 3.12, (1) implies (2). Further, (2) implies (3) is just the one-dimensional case of Theorem 2.12, and an independent proof follows easily (under the present assumption (H1)—(H3) on h) from the proof of Theorem 3.10 (see Remark 3.11). Moreover, (3) implies (1) follows from Remark 3.7, while statements (3) and (4) are clearly equivalent. ∎

We now briefly explore more relations among the one-sided lower estimates. It turns out that the natural analogue of Theorem 3.13 is not true.

Theorem 3.14 *Let $h \in G_d$ and $(l_j)_{j=1}^{\infty}$ be an arbitrary nonincreasing positive sequence with $\sum_{j=1}^{\infty} l_j < \infty$. Then $\alpha > 0$ implies $M_* > 0$ and $\delta_* > 0$ implies $M_* > 0$.*

Proof. In view of definitions (3.20) and (3.23), the first assertion follows from the inequality in (3.34). The second assertion follows immediately by taking the lower limit as $x \to \infty$ in (3.67). ∎

As is shown in [LaPo2] for the standard case when $h(x) = x^d$, there are no other relationships between α, M_* and δ_* than those given in the above theorem. (See [LaPo2, Examples 3.12-3.14, p. 56].)

4 Spectra of Fractal Strings and the Riemann Zeta-Function

We still work here, as in Sections 3 and 2.1, in the one-dimensional case (i.e., when $n = 1$); further, as far as possible, we will continue to use the notation from those sections. We will prove Theorem 2.5 (resp., 2.4) in Section 4.1 (resp., 4.2) below. Corollary 2.6 will then follow by combining Theorems 2.4 and 2.5.

4.1 Characterization of h-Minkowski measurability

Let $(l_j)_{j=1}^{\infty}$ be a nonincreasing sequence of positive numbers such that $\sum_{j=1}^{\infty} l_j < \infty$ and let $L(h; \epsilon)$ be given by (3.20). We have shown in Theorem 3.4(b) that if $l_j \sim Lg(j)$ for some $L > 0$, then the sequence $(l_j)_{j=1}^{\infty}$ is h-Minkowski measurable (that is, $\lim_{\epsilon \to 0+} L(h; \epsilon)$ exists in $(0, \infty)$) with h-Minkowski content given by $M := \lim_{\epsilon \to 0+} L(h; \epsilon) = \frac{2^{1-d}}{1-d} L^d$. We establish here the converse of this result and hence characterize the situation when $(l_j)_{j=1}^{\infty}$ is h-Minkowski measurable with $h \in G_d$. Theorem 2.5 will then follow from Theorem 3.4(b) and Theorem 4.1 below.

Theorem 4.1 *Suppose that $(l_j)_{j=1}^{\infty}$ is a nonincreasing sequence of positive numbers with $\sum_{j=1}^{\infty} l_j < \infty$ and such that for some $L > 0$ and $h \in G_d$, the following limit exists and is given by*

$$\lim_{\epsilon \to 0+} L(h; \epsilon) = \frac{2^{1-d}}{1-d} L^d. \tag{4.1}$$

Then we have

$$l_j \sim Lg(j), \quad as \quad j \to \infty. \tag{4.2}$$

(Recall that $g(x) = h^{-1}(1/x)$.)

38

Proof. The proof of this theorem is divided into two steps. The key part of the proof is contained in Step 1. It establishes the result in the case when the sequence $(l_j)_{j=1}^\infty$ is strictly decreasing and $l_{j+1}/l_j \to 1$ as $j \to \infty$. Then in Step 2, we will show that the general case can always be reduced to this special case.

Step 1.   We assume now that $(l_j)_{j=1}^\infty$ is strictly decreasing and $l_{j+1}/l_j \to 1$ as $j \to \infty$.

For each $j \geq 1$, define α_j, β_j by the equations:

$$l_j = \alpha_j g(j) \quad \text{and} \quad \sum_{k>j} l_k = \beta_j \int_j^\infty g(x)dx. \tag{4.3}$$

Note that since we assume that $M_* = M^* = M = \frac{2^{1-d}}{1-d}L^d$, we have by Theorem 2.7 that $0 < \alpha \; (= \liminf_{\epsilon \to 0^+} \alpha_j) \leq \beta \; (= \limsup_{\epsilon \to 0^+} \alpha_j) < \infty$. Thus the sequence $(\alpha_j)_{j=1}^\infty$ is contained in some compact subset of $(0, \infty)$. Since $(l_j)_{j=1}^\infty$ is strictly decreasing, we have $J(l_j) = j$ for all $j \geq 1$. (Recall that $J(\epsilon)$ is defined by (3.19).) Then, in view of (3.20),

$$
\begin{aligned}
L(h; \tfrac{1}{2}l_j) &= 2h(\tfrac{1}{2}l_j)J(l_j) + \frac{2}{l_j}h(\tfrac{1}{2}l_j) \sum_{k>J(l_j)} l_k \\
&= 2h(\tfrac{\alpha_j}{2}g(j))j + \frac{2}{l_j}h(\tfrac{\alpha_j}{2}g(j)) \sum_{k>J(l_j)} l_k \\
&= 2(\tfrac{\alpha_j}{2})^d(1 + o(1))h(g(j))j \\
&\quad + \frac{2}{\alpha_j g(j)}(\tfrac{\alpha_j}{2})^d(1 + o(1))h(g(j)) \sum_{k>j} l_k, \tag{4.4}
\end{aligned}
$$

where $o(1) \to 0$ as $j \to \infty$. The last equality follows from Proposition 3.2(i). Since $h(g(j)) = h(h^{-1}(1/j)) = 1/j$, we see that

$$
\begin{aligned}
L(h; \tfrac{1}{2}l_j) &= 2^{1-d}\alpha_j^d(1 + o(1)) \\
&\quad + 2^{1-d}\alpha_j^{d-1}(1 + o(1))\frac{1}{jg(j)}\beta_j \int_j^\infty g(x)dx. \tag{4.5}
\end{aligned}
$$

We know that $l_j \to 0$ and, by Lemma 3.3(1), $\frac{\int_j^\infty g(x)dx}{jg(j)} \to \frac{d}{1-d}$ as $j \to \infty$. So if we let $j \to \infty$ in (4.5), we deduce from hypothesis (4.1) that

$$\lim_{j \to \infty} L(h; \tfrac{l_j}{2}) = \lim_{j \to \infty} (2^{1-d}\alpha_j^d + 2^{1-d}\alpha_j^{d-1}\beta_j \frac{d}{1-d}) = \frac{2^{1-d}}{1-d}L^d, \tag{4.6}$$

or, equivalently,

$$\lim_{j \to \infty} (\alpha_j^d + \alpha_j^{d-1} \beta_j \frac{d}{1-d}) = \frac{L^d}{1-d}. \tag{4.7}$$

The first equality in (4.6) holds because $(\alpha_j)_{j=1}^\infty$ is contained in a compact subset of $(0, \infty)$, and so $\alpha_j^d o(1) = o(1)$. Also,

$$\sum_{k>j} l_k = \beta_j \int_j^\infty g(x)dx = \sum_{k>j} \alpha_k g(k)$$

$$\leq \beta(1 + o(1)) \sum_{k>j} g(k) \leq \beta(1 + o(1)) \int_j^\infty g(x)dx,$$

for all j large enough. Thus $\beta_j < \beta(1 + o(1))$ for all j large enough. Then we also have $\beta_j o(1) = o(1)$. So we can deduce (4.6) from (4.5).

We now want to show that

$$\limsup_{j \to \infty} \beta_j \leq L. \tag{4.8}$$

Define the positive number γ_j by the equation

$$\alpha_j^d + \alpha_j^{d-1} \beta_j \frac{d}{1-d} = \frac{\gamma_j^d}{1-d}.$$

From (4.7), $\gamma_j \to L$ as $j \to \infty$. For fixed $b > 0$ and $d \in (0, 1)$, the minimum value of $a^d + a^{d-1} b \frac{d}{1-d}$ on the interval $(0, \infty)$ occurs when $a = b$ and is equal to $\frac{b^d}{1-d}$. Thus

$$\beta_j^d \frac{1}{1-d} \leq \gamma_j^d \frac{1}{1-d};$$

so $\beta_j \leq \gamma_j$ for each j, which proves (4.8).

Next we will show that

$$\liminf_{j \to \infty} \alpha_j \leq L \quad \text{and} \quad \limsup_{j \to \infty} \alpha_j \geq L. \tag{4.9}$$

If the first inequality in (4.9) fails, then there is some $\theta > 0$ with $\alpha_j > L + \theta$ for all large j. Then for j large enough,

$$\beta_j \int_j^\infty g(x)dx = \sum_{k>j} l_k = \sum_{k>j} \alpha_k g(k)$$

$$> (L + \theta) \sum_{k>j} g(k) > (L + \theta) \int_{j+1}^\infty g(x)dx.$$

The last inequality holds since g is a decreasing function. So we have

$$\beta_j > (L+\theta)\frac{\int_{j+1}^{\infty} g(x)dx}{\int_j^{\infty} g(x)dx}, \tag{4.10}$$

and since the ratio of the two integrals on the right side of (4.10) tends to 1 as $j \to \infty$ by Proposition 3.2(iv), we see by letting $j \to \infty$ in (4.10) that

$$\liminf_{j\to\infty} \beta_j \geq L+\theta,$$

which contradicts (4.8). Therefore, $\liminf_{j\to\infty} \alpha_j \leq L$.

To prove the second inequality in (4.9), let $\eta := \limsup_{j\to\infty} \alpha_j$. Then an argument analogous to the one just completed above shows that

$$\limsup_{j\to\infty} \beta_j \leq \limsup_{j\to\infty} \alpha_j = \eta.$$

Let $j_1 \leq j_2 \leq \cdots$ be a sequence of natural numbers for which $\lim_{k\to\infty} \alpha_{j_k} = \eta$. Then $\beta_{j_k} \leq \alpha_{j_k} + o(1)$ as $k \to \infty$. Thus

$$\frac{\gamma_{j_k}^d}{1-d} = \alpha_{j_k}^d + \alpha_{j_k}^{d-1}\beta_{j_k}\frac{d}{1-d}$$

$$\leq \alpha_{j_k}^d + \alpha_{j_k}^d\frac{d}{1-d} + \frac{d}{1-d}\alpha_{j_k}^{d-1}o(1) = \frac{\alpha_{j_k}^d}{1-d} + o(1).$$

Since $\gamma_{j_k} \to L$ and $\alpha_{j_k} \to \eta$ as $k \to \infty$, we deduce from the above inequality that

$$\frac{L^d}{1-d} \leq \frac{\eta^d}{1-d},$$

which shows that $L \leq \eta = \limsup_{j\to\infty} \alpha_j$ and establishes (4.9).

We now prove that $\limsup_{j\to\infty} \alpha_j = L$. Suppose $\limsup_{j\to\infty} \alpha_j > L$. Then there is some $\theta > 0$ and an infinite set of subscripts with $\alpha_j > L+\theta$. Let

$$r_\theta = \left(\frac{L+\frac{1}{2}\theta}{L+\theta}\right)^d, \quad \text{so } 0 < r_\theta < 1.$$

If j_1 is large with $\alpha_{j_1} > L+\theta$, let $j_0 = [r_\theta j_1]$ and let j_2 be the least subscript greater than j_1 with $\alpha_{j_2} < L+1/j_1$. By the first inequality in (4.9), such j_2 exists. By assumption, $l_{j+1}/l_j \to 1$; that is, $\frac{\alpha_{j+1}g(j+1)}{\alpha_j g(j)} \to 1$ as $j \to \infty$.

Further, by Proposition 3.2(ii), $\frac{g(j+1)}{g(j)} \to 1$. Hence it follows that $\frac{\alpha_{j+1}}{\alpha_j} \to 1$ as $j \to \infty$, which implies that $\alpha_{j_2} \to L$ as $j_1 \to \infty$ through those numbers with $\alpha_{j_1} > L + \theta$. Thus by (4.7), $\beta_{j_2} \to L$.

If $j_0 + 1 \leq j < j_1$, then $l_j > l_{j_1}$. So for j_1 large enough,

$$
\begin{aligned}
\alpha_j &> \alpha_{j_1} \frac{g(j_1)}{g(j)} = \alpha_{j_1} \frac{g(\frac{j_1}{j}j)}{g(j)} \\
&\geq \alpha_{j_1} \frac{g(\frac{1}{r_\theta}j)}{g(j)} = \alpha_{j_1} r_\theta^{1/d}(1 + o(1)) \\
&= \alpha_{j_1} \frac{L + \theta/2}{L + \theta}(1 + o(1)) > (L + \frac{\theta}{2})(1 + o(1)),
\end{aligned}
\tag{4.11}
$$

where $o(1) \to 0$ as $j_1 \to \infty$. (Since $j_1/j \leq 1/r_\theta$, for $j_0 + 1 \leq j < j_1$, and g is decreasing, we have $g(j_1) \geq g(\frac{j}{r_\theta})$. Moreover, in the second equality above, we used again Proposition 3.2(ii).)

Further, if $j_1 < j < j_2$, then $\alpha_j > L$. Hence

$$
\begin{aligned}
\beta_{j_0} \int_{j_0}^\infty g(x)dx &= \sum_{j>j_0} l_j = \sum_{j>j_0} \alpha_j g(j) \\
&= \left(\sum_{j=j_0+1}^{j_1} + \sum_{j=j_1+1}^{j_2-1} + \sum_{j=j_2}^\infty \right) \alpha_j g(j) \\
&> (L + \frac{\theta}{2})(1 + o(1)) \int_{j_0+1}^{j_1+1} g(x)dx \\
&\quad + L \int_{j_1+1}^{j_2} g(x)dx + \beta_{j_2} \int_{j_2}^\infty g(x)dx \\
&> L \int_{j_0+1}^{j_2} g(x)dx + \frac{\theta}{2}(1 + o(1)) \int_{j_0+1}^{j_1} g(x)dx \\
&\quad + \beta_{j_2} \int_{j_2}^\infty g(x)dx.
\end{aligned}
\tag{4.12}
$$

Therefore,

$$
\begin{aligned}
\beta_{j_0} &> L \frac{\int_{j_0+1}^{j_2} g(x)dx}{\int_{j_0}^\infty g(x)dx} + \frac{\theta}{2}(1 + o(1)) \frac{\int_{j_0+1}^{j_1} g(x)dx}{\int_{j_0}^\infty g(x)dx} + \beta_{j_2} \frac{\int_{j_2}^\infty g(x)dx}{\int_{j_0}^\infty g(x)dx} \\
&= L \frac{(\int_{j_0+1}^\infty - \int_{j_2}^\infty)g(x)dx}{\int_{j_0}^\infty g(x)dx} + \frac{\theta}{2}(1 + o(1)) \frac{(\int_{j_0+1}^\infty - \int_{j_1}^\infty)g(x)dx}{\int_{j_0}^\infty g(x)dx}
\end{aligned}
$$

$$+\beta_{j_2}\frac{\int_{j_2}^{\infty}g(x)dx}{\int_{j_0}^{\infty}g(x)dx}$$

$$= (\beta_{j_2}-L)\frac{\int_{j_2}^{\infty}g(x)dx}{\int_{j_0}^{\infty}g(x)dx} + L\frac{\int_{j_0+1}^{\infty}g(x)dx}{\int_{j_0}^{\infty}g(x)dx}$$

$$+\frac{\theta}{2}(1+o(1))\frac{(\int_{j_0+1}^{\infty}-\int_{j_1}^{\infty})g(x)dx}{\int_{j_0}^{\infty}g(x)dx}$$

$$=: \quad A+B+C, \text{ say.} \tag{4.13}$$

Since $j_2 > j_1 > j_0$, we have $(\int_{j_2}^{\infty}g(x)dx/\int_{j_0}^{\infty}g(x)dx) < 1$. Thus

$$A := (\beta_{j_2}-L)\frac{\int_{j_2}^{\infty}g(x)dx}{\int_{j_0}^{\infty}g(x)dx}$$

satisfies $|A| \leq |\beta_{j_2}-L|$. We showed before that $\beta_{j_2} \to L$ as $j_1 \to \infty$; so we get $A \to 0$ as $j_1 \to \infty$, which forces $j_2 \to \infty$.

Next

$$B := L\frac{\int_{j_0+1}^{\infty}g(x)dx}{\int_{j_0}^{\infty}g(x)dx} \to L \quad \text{as} \quad j_0 \to \infty,$$

by Proposition 3.2(iv). (This also follows from the Cauchy criterion for convergent improper integrals.)

Finally, consider

$$C := \frac{\theta}{2}(1+o(1))\left(\frac{\int_{j_0+1}^{\infty}g(x)dx}{\int_{j_0}^{\infty}g(x)dx} - \frac{\int_{j_1}^{\infty}g(x)dx}{\int_{j_0}^{\infty}g(x)dx}\right).$$

The first ratio tends to 1 as $j_0 \to \infty$. As for the second ratio, using Proposition 3.2(iv) and the fact that $j_1 \sim (1/r_\theta)j_0$ as $j_1 \to \infty$, we see that it tends to $r_\theta^{-(1-1/d)}$ as $j_1 \to \infty$, which forces $j_0 \to \infty$; therefore,

$$C \to \frac{\theta}{2}(1-r_\theta^{-(1-1/d)}) \quad \text{as} \quad j_0 \to \infty.$$

By taking the upper limit of (4.13) as $j_0 \to \infty$, we obtain that (since $1 - r_\theta^{\frac{1}{d}-1} > 0$)

$$\limsup_{j_0\to\infty} \beta_{j_0} \geq L + \frac{\theta}{2}(1-r_\theta^{1/d-1}) > L,$$

which contradicts (4.8). Thus we deduce that

$$\limsup_{j\to\infty} \alpha_j = L.$$

To complete the proof of Step 1, we must show that $\liminf_{j\to\infty} \alpha_j \geq L$. Suppose not. Then there are some θ with $0 < \theta < 1$ and an infinite set S of subscripts j with $\alpha_j \leq L(1-\theta)$ for all $j \in S$. Let

$$\epsilon = \theta - \frac{1-\theta}{d}((1-\theta)^{-d} - 1).$$

Since $(1-\theta)^{1-d} < 1 - (1-d)\theta$ and $(1-\theta)^{-d} > 1$, it easily follows that $0 < \epsilon < \theta$.

For each large $j_1 \in S$, there is a unique $j_0 = j_0(j_1)$ such that

$$j_0 < j_1, \ \alpha_{j_0} > L(1-\epsilon) \ \text{ and if } \ j_0 < j \leq j_1, \text{ then } \alpha_j \leq L(1-\epsilon).$$

Indeed, the existence of j_0 for all large $j_1 \in S$ follows from the second inequality in (4.9). For any $j_2 > j_1$, we have

$$
\begin{aligned}
\beta_{j_0} \int_{j_0}^{\infty} g(x)dx &= \sum_{j > j_0} l_j = \left(\sum_{j=j_0+1}^{j_1} + \sum_{j=j_1+1}^{j_2} + \sum_{j=j_2+1}^{\infty} \right) \alpha_j g(j) \\
&\leq L(1-\epsilon) \int_{j_0}^{j_1} g(x)dx + (j_2 - j_1)l_{j_1} + \beta_{j_2} \int_{j_2}^{\infty} g(x)dx \\
&= L(1-\epsilon) \int_{j_0}^{j_1} g(x)dx + (j_2 - j_1)\alpha_{j_1}g(j_1) + \beta_{j_2} \int_{j_2}^{\infty} g(x)dx \\
&\leq L(1-\epsilon) \int_{j_0}^{j_1} g(x)dx + (j_2 - j_1)L(1-\theta)g(j_1) \\
&\quad + \beta_{j_2} \int_{j_2}^{\infty} g(x)dx. \hspace{2cm} (4.14)
\end{aligned}
$$

Thus we obtain

$$
\begin{aligned}
\beta_{j_0} &\leq L(1-\epsilon)\frac{(\int_{j_0}^{\infty} - \int_{j_1}^{\infty})g(x)dx}{\int_{j_0}^{\infty} g(x)dx} + L(1-\theta)\frac{(\frac{j_2}{j_1} - 1)j_1 g(j_1)}{\int_{j_0}^{\infty} g(x)dx} + \beta_{j_2}\frac{\int_{j_2}^{\infty} g(x)dx}{\int_{j_0}^{\infty} g(x)dx} \\
&= L(1-\epsilon)\left(1 - \frac{\int_{j_1}^{\infty} g(x)dx}{\int_{j_0}^{\infty} g(x)dx}\right) + L(1-\theta)(\frac{j_2}{j_1} - 1)\frac{j_1 g(j_1)}{\int_{j_1}^{\infty} g(x)dx}\frac{\int_{j_1}^{\infty} g(x)dx}{\int_{j_0}^{\infty} g(x)dx} \\
&\quad + \beta_{j_2}\frac{\int_{j_2}^{\infty} g(x)dx}{\int_{j_1}^{\infty} g(x)dx}\frac{\int_{j_1}^{\infty} g(x)dx}{\int_{j_0}^{\infty} g(x)dx}, \hspace{2cm} (4.15)
\end{aligned}
$$

and this is true for all $j_2 > j_1$.

By Lemma 3.3(1), we know that $\frac{j_1 g(j_1)}{\int_{j_1}^\infty g(x)dx} \to \frac{1-d}{d}$ as $j_1 \to \infty$. Now if we choose $j_2 = [(1-\theta)^{-d}j_1]$, which implies that $j_2 \sim (1-\theta)^{-d}j_1$ as $j_1 \to \infty$, then by Proposition 3.2(iv), we have

$$\frac{\int_{j_2}^\infty g(x)dx}{\int_{j_1}^\infty g(x)dx} \to (1-\theta)^{-d(1-1/d)} = (1-\theta)^{1-d}, \quad \text{as } j_1 \to \infty.$$

If we put these into (4.15), we will obtain

$$\beta_{j_0} \leq L(1-\epsilon) - L(1-\epsilon)\frac{\int_{j_1}^\infty g(x)dx}{\int_{j_0}^\infty g(x)dx} + L(1-\theta)$$

$$\cdot\left((1-\theta)^{-d}(1+o(1)) - 1\right)\frac{1-d}{d}(1+o(1))\frac{\int_{j_1}^\infty g(x)dx}{\int_{j_0}^\infty g(x)dx}$$

$$+ L(1+o(1))(1-\theta)^{1-d}\frac{\int_{j_1}^\infty g(x)dx}{\int_{j_0}^\infty g(x)dx}$$

$$\text{(we have } \beta_{j_2} \leq L(1+o(1)) \text{ since } \limsup_{j\to\infty} \beta_j \leq L)$$

$$= L(1-\epsilon) + L(1+o(1))\frac{\int_{j_1}^\infty g(x)dx}{\int_{j_0}^\infty g(x)dx}$$

$$\cdot\left(-(1-\epsilon) + (1-\theta)((1-\theta)^{-d} - 1)\frac{1-d}{d} + (1-\theta)^{1-d} + o(1)\right),$$

$$(4.16)$$

where $o(1)$ tends to 0 as $j_1 \to \infty$ through S.

The large parenthetical expression in (4.16), apart from the term involving $o(1)$, is

$$-1+\epsilon+(1-\theta)\left(\frac{(1-\theta)^{-d}}{d} + \frac{1-d}{d}\right) = -\theta + \epsilon + \frac{1-\theta}{d}((1-\theta)^{-d} - 1) = 0,$$

by our choice of ϵ. Thus we deduce that

$$\beta_{j_0} \leq L(1-\epsilon) + o(1).$$

From the definition of j_0 and the assumption $l_{j+1}/l_j \to 1$, we have $\alpha_{j_0} = L(1-\epsilon) + o(1)$. Consequently,

$$\frac{\gamma_{j_0}^d}{1-d} = \alpha_{j_0}^d + \alpha_{j_0}^{d-1}\beta_{j_0}\frac{d}{1-d} \leq \frac{L^d(1-\epsilon)^d}{1-d} + o(1) < \frac{L^d}{1-d},$$

as $j_1 \to \infty$ through S, contradicting (4.7). Hence S must be a finite set and we deduce that $\liminf_{j\to\infty} \alpha_j \geq L$. In conjunction with (4.9), this shows that $\lim_{j\to\infty} \alpha_j = L$ (i.e., $l_j \sim Lg(j)$ as $j \to \infty$) and thus concludes the proof of Step 1.

Step 2. We now show that we can always reduce the general case to the situation when $(l_j)_{j=1}^\infty$ is strictly decreasing and $l_{j+1}/l_j \to 1$ as $j \to \infty$.

Let (l_j) be an arbitrary sequence satisfying the hypothesis of the theorem. If $l_j > l_{j+1}$, we have $J(l_j) = j$ and

$$2h(\frac{l_j}{2})J(l_j) = 2h(\frac{l_j}{2})j = 2h(\frac{\alpha_j}{2}g(j))j = 2^{1-d}\alpha_j^d(1 + o(1)).$$

Since $2h(l_j/2)J(l_j) \leq L(h; l_j/2)$ and we assume (4.1), then

$$2^{1-d}\alpha_j^d(1 + o(1)) \leq L(h; \frac{l_j}{2}) = \frac{2^{1-d}L^d}{1-d}.$$

(Recall that $l_j \to 0$ as $j \to \infty$.)

Thus

$$\limsup_{j:l_j>l_{j+1}} \alpha_j \leq L(1-d)^{-1/d}.$$

If $l_j = l_{j+1} = \cdots = l_{j_0} > l_{j_0+1}$, then $\alpha_j < \alpha_{j+1} < \cdots < \alpha_{j_0}$, so that

$$\limsup_{j\to\infty} \alpha_j \leq L(1-d)^{-1/d}. \tag{4.17}$$

We now show that the hypothesis of the theorem forces $\lim_{j\to\infty} \frac{l_{j+1}}{l_j} = 1$, which is one of the two facts we need to complete the proof.

Suppose $l_j \geq \epsilon > l_{j+1}$. Then $J(\epsilon) = j$. Write $\epsilon = rl_j$, where $1 \geq r > \frac{l_{j+1}}{l_j}$, and let $c = \liminf_{j\to\infty} \frac{l_{j+1}}{l_j}$. We claim that $c > 0$. If not, suppose $(j_k)_{k=1}^\infty$ is a sequence such that $\lim_{k\to\infty} l_{j_k+1}/l_{j_k} = c = 0$; that is,

$$\frac{\alpha_{j_k+1}g(j_k+1)}{\alpha_{j_k}g(j_k)} \to 0 \quad \text{as} \quad k \to \infty.$$

Since $g(j_k+1)/g(j_k) \to 1$ by Proposition 3.2, this forces $\alpha_{j_k+1}/\alpha_{j_k} \to 0$ as $k \to \infty$, which is impossible since we know from (4.1) and Theorem 2.7

that $0 < \alpha \ (= \liminf_{j \to \infty} \alpha_j) \leq \beta \ (= \limsup_{j \to \infty} \alpha_j) < \infty$. Therefore $c > 0$. Thus, for all j with $l_{j+1} < l_j$, r is contained in some compact subset of $(0, \infty)$.

Then we see that for $l_{j+1} < \epsilon \leq l_j$:

$$
\begin{aligned}
L(h; \frac{\epsilon}{2}) &= 2h(\frac{\epsilon}{2})J(\epsilon) + (\frac{\epsilon}{2})^{-1}h(\frac{\epsilon}{2}) \sum_{k > J(\epsilon)} l_k \\
&= 2h(\frac{r}{2}l_j)j + (\frac{rl_j}{2})^{-1}h(\frac{r}{2}l_j) \sum_{k > j} l_k \\
&= h(\frac{r}{2}l_j)\left(2j + (\frac{rl_j}{2})^{-1} \sum_{k > j} l_k\right) \\
&= r^d(1 + o(1))h(\frac{l_j}{2})\left(2j + (\frac{rl_j}{2})^{-1} \sum_{k > j} l_k\right) \\
&=: (r^d A_j + r^{d-1} B_j)(1 + o(1)),
\end{aligned}
\tag{4.18}
$$

where

$$
A_j := 2h(\frac{l_j}{2})j = 2h(\frac{l_j}{2})J(l_j), \quad B_j := (\frac{l_j}{2})^{-1}h(\frac{l_j}{2}) \sum_{k > j} l_k.
$$

So we can write

$$
L(h; \frac{\epsilon}{2}) = (r^d A_j + r^{d-1} B_j)(1 + o(1)),
$$

where '$o(1)$' tends to 0 as $j \to \infty$. By (3.20), $A_j + B_j = L(h; l_j/2)$, which tends to $\frac{2^{1-d}}{1-d}L^d$ as $j \to \infty$ through numbers with $l_j > l_{j+1}$.

Since each A_j, B_j is positive, the set $\{(A_j, B_j) : l_j > l_{j+1}\}$ is then contained in some compact subset K of $\mathbf{R}^2 - \{(0, 0)\}$.

Suppose that there is some $\theta \in (0, 1)$ such that $l_{j+1}/l_j < 1 - \theta$ for infinitely many j. For any fixed A, B, let $\omega(A, B)$ denote the difference between the maximum and minimum values of the function $r^d A + r^{d-1} B$ on the compact interval $[1 - \theta, 1]$. It can be easily checked that $\omega(A, B)$ is continuous and positive on $\mathbf{R}^2 - \{(0, 0)\}$. Hence it assumes a positive minimum κ on the compact set K. So for each $(A, B) \in K$, there exist $r_1, r_2 \in [1 - \theta, 1]$ such that

$$
\omega(A, B) = r_1^d A + r_1^{d-1} B - (r_2^d A + r_2^{d-1} B) \geq \kappa.
\tag{4.19}
$$

Therefore, for each j with $l_{j+1}/l_j < 1 - \theta$, there are $r_1, r_2 \in [1-\theta, 1]$ such that (4.19) holds. Thus

$$L(h; \frac{r_1 l_j}{2}) - L(h; \frac{r_2 l_j}{2}) = \omega(A_j, B_j)(1 + o(1)) \geq \kappa(1 + o(1)).$$

This contradicts (4.1), so we have $\lim_{j\to\infty} l_{j+1}/l_j = 1$.

For any real number $\epsilon > 0$ and $j \in \mathbf{N}$, define $L_h(\epsilon, j)$ by

$$L_h(\epsilon, j) := 2h(\epsilon)j + \epsilon^{-1}h(\epsilon) \sum_{k>j} l_k.$$

For example, by (3.20), $L_h(\epsilon, J(2\epsilon)) = L(h; \epsilon)$. We now show that if j_ϵ is any integer with $l_{j_\epsilon} = l_{J(2\epsilon)}$, then no matter how j_ϵ is chosen for each $\epsilon > 0$, we have

$$\lim_{\epsilon\to 0^+} L_h(\epsilon, j_\epsilon) = \lim_{\epsilon\to 0^+} L(h; \epsilon). \tag{4.20}$$

Given $\epsilon > 0$, write $j = j_\epsilon$ and $j_0 = J(2\epsilon)$. Then

$$\begin{aligned} L_h(\epsilon, j) - L(h; \epsilon) &= 2h(\epsilon)(j - j_0) + \epsilon^{-1}h(\epsilon)\sum_{k=j+1}^{j_0} l_k \\ &= 2h(\epsilon)(j - j_0) + \epsilon^{-1}h(\epsilon)(j_0 - j)l_{j_0} \\ &= \epsilon^{-1}h(\epsilon)(l_{j_0} - 2\epsilon)(j_0 - j). \end{aligned} \tag{4.21}$$

The second equality is a consequence of the following fact: $j_0 \geq j$ means $l_{j_0} \leq l_j$, but we have $l_j = l_{j_0}$; thus for all $j \leq k \leq j_0$, we have $l_j = l_{j+1} = \cdots = l_k = \cdots = l_{j_0}$. So continuing with (4.21), we see that

$$\begin{aligned} 0 &\leq L_h(\epsilon, j) - L(h; \epsilon) = \epsilon^{-1}h(\epsilon)(l_{j_0} - 2\epsilon)(j_0 - j) \\ &\leq \epsilon^{-1}h(\epsilon)j_0(l_{j_0} - 2\epsilon). \end{aligned}$$

Since $j_0 = J(2\epsilon)$, which means $l_{j_0} \geq 2\epsilon > l_{j_0+1}$, we have

$$\begin{aligned} 0 \leq L_h(\epsilon, j) - L(h; \epsilon) &\leq (\frac{l_{j_0+1}}{2})^{-1}h(\frac{l_{j_0}}{2})j_0(l_{j_0} - l_{j_0+1}) \\ &= (\frac{l_{j_0+1}}{2})^{-1}h(\frac{\alpha_{j_0}}{2}g(j_0))j_0(l_{j_0} - l_{j_0+1}) \\ &= 2(\frac{\alpha_{j_0}}{2})^d(1 + o(1))\frac{1}{j_0}j_0(\frac{l_{j_0}}{l_{j_0+1}} - 1) \\ &= 2(\frac{\alpha_{j_0}}{2})^d(1 + o(1))(\frac{l_{j_0}}{l_{j_0+1}} - 1). \end{aligned} \tag{4.22}$$

So if we let $\epsilon \to 0^+$ in (4.22) (which forces $j_0 \to \infty$) and note that (4.17) holds, we obtain (4.20), because as we have seen above, $l_{j_0}/l_{j_0+1} \to 1$.

We now define a sequence $(m_j)_{j=1}^\infty$ of nonnegative reals with the following properties. We let $m_1 = 0$. If $l_j > l_{j+1}$, we let $m_j = 0$. If $l_j = l_{j+1} = \cdots = l_{j+k} > l_{j+k+1}$, we choose $0 < m_{j+1} < m_{j+2} < \cdots < m_{j+k} < l_{j+k+1} - l_{j+k}$. Further, we choose the positive m values so small that

$$\sum_{q>j} m_q = o\left(\frac{l_j}{h(l_j)}\right) \quad \text{and} \quad m_j = o(l_j) \quad \text{as} \quad j \to \infty. \tag{4.23}$$

Then the positive sequence $(l_j - m_j)_{j=1}^\infty$ is strictly decreasing and $l_j \sim Lg(j)$ if and only if $l_j - m_j \sim Lg(j)$, as $j \to \infty$. (This follows since by (4.1) and Theorem 2.7, $l_j \asymp g(j)$ and hence in particular, $o(l_j)/g(j) = o(1)$, as $j \to \infty$.)

We now show that $(l_j - m_j)_{j=1}^\infty$ satisfies the same hypothesis (4.1) as $(l_j)_{j=1}^\infty$, thus completing the proof of Step 2.

Let $J'(\epsilon)$ be the analogue of the function $J(\epsilon)$ for the sequence $(l_j - m_j)_{j=1}^\infty$. We need to prove that

$$L'(h;\epsilon) := 2h(\epsilon)J'(2\epsilon) + \epsilon^{-1}h(\epsilon) \sum_{j>J'(2\epsilon)} (l_j - m_j)$$

satisfies

$$L'(h;\epsilon) \to \frac{2^{1-d}}{1-d}L^d, \quad \text{as} \quad \epsilon \to 0^+.$$

To this end, it will suffice to show that $L'(h;\epsilon) - L(h;\epsilon) \to 0$.

From the above construction, we have $l_{J(\epsilon)} = l_{J'(\epsilon)}$ for any $\epsilon > 0$. Indeed, since $l_j - m_j \le l_j$, we have $J'(\epsilon) \le J(\epsilon)$. If $J'(\epsilon) < J(\epsilon)$, then since $l_{J'(\epsilon)} - m_{J'(\epsilon)} \ge \epsilon > l_j - m_j$ for all $j > J'(\epsilon)$ and since $l_{J(\epsilon)} \ge \epsilon$, we have $m_j > 0$ for all $j = J'(\epsilon) + 1, \ldots, J(\epsilon)$. Thus $l_{J(\epsilon)} = l_{J'(\epsilon)}$. We write

$$L'(h;\epsilon) - L(h;\epsilon) = (L'(h;\epsilon) - L_h(\epsilon, J'(2\epsilon))) + (L_h(\epsilon, J'(2\epsilon)) - L(h;\epsilon)). \tag{4.24}$$

Now, by (4.20), $L_h(\epsilon, J'(2\epsilon)) - L(h;\epsilon) \to 0$ as $\epsilon \to 0^+$. Further,

$$\begin{aligned}
|L'(h;\epsilon) - L_h(\epsilon, J'(2\epsilon))| &= |2h(\epsilon)J'(2\epsilon) + \epsilon^{-1}h(\epsilon) \sum_{j>J'(2\epsilon)} (l_j - m_j) \\
&\quad - 2h(\epsilon)J'(2\epsilon) - \epsilon^{-1}h(\epsilon) \sum_{j>J'(2\epsilon)} l_j|
\end{aligned}$$

$$= \epsilon^{-1}h(\epsilon) \sum_{j>J'(2\epsilon)} m_j$$

$$= \epsilon^{-1}h(\epsilon)\, o\left(\frac{l_{J'(2\epsilon)}}{h(l_{J'(2\epsilon)})}\right). \qquad (4.25)$$

We know that $l_{J'(2\epsilon)} = l_{J(2\epsilon)}$ for all $\epsilon > 0$ and

$$1 \le \frac{l_{J(\epsilon)}}{\epsilon} < \frac{l_{J(\epsilon)}}{l_{J(\epsilon)+1}} \to 1, \quad \text{as} \ \ \epsilon \to 0^+.$$

So we see that $l_{J(\epsilon)}/\epsilon \to 1$ as $\epsilon \to 0^+$. Then we deduce that

$$\lim_{\epsilon \to 0^+} \epsilon^{-1}h(\epsilon)\frac{l_{J(2\epsilon)}}{h(l_{J(2\epsilon)})} = \lim_{\epsilon \to 0^+} 2\frac{l_{J(2\epsilon)}}{2\epsilon}\frac{h(2\epsilon)}{h(l_{J(2\epsilon)})}\frac{h(\epsilon)}{h(2\epsilon)}$$

$$= 2 \cdot 1 \cdot 1 \cdot 2^{-d} = 2^{1-d}.$$

Hence

$$\epsilon^{-1}h(\epsilon)\, o\left(\frac{l_{J'(2\epsilon)}}{h(l_{J'(2\epsilon)})}\right) = o(1), \quad \text{as} \ \ \epsilon \to 0^+.$$

So from (4.25), we have $\lim_{\epsilon \to 0^+}(L'(h;\epsilon) - L_h(\epsilon; J'(2\epsilon))) = 0$. Thus, using (4.24), we obtain that

$$\lim_{\epsilon \to 0^+}(L(h;\epsilon) - L'(h;\epsilon)) = 0.$$

Therefore

$$\lim_{\epsilon \to 0^+} L'(h;\epsilon) = \frac{2^{1-d}}{1-d}L^d.$$

Since the sequence $(l_j - m_j)_{j=1}^\infty$ is strictly decreasing, it follows from Step 1 that $l_j - m_j \sim Lg(j)$ as $j \to \infty$ and hence also $l_j \sim Lg(j)$ as $j \to \infty$. (According to a previous argument, we necessarily have $\frac{l_{j+1}-m_{j+1}}{l_j-m_j} \to 1$. This can also be checked directly since $m_j = o(l_j)$ and $\frac{l_{j+1}}{l_j} \to 1$.)

We have now completed the proof of Step 2, and thus of Theorem 4.1. ∎

Remark 4.2 The present proof follows the lines of [LaPo2, Theorem 4.1], even though it is more intricate. Recently, In [Fa3], Falconer has recently given a 'dynamical systems proof' of the characterization of Minkowski measurability obtained in [LaPo1,2]. In the present more general situation, it might be interesting to provide a proof of Theorem 4.1 along those lines.

4.2 Existence of a monotonic second term: the Riemann zeta-function

We conclude the proof of Corollary 2.6 by proving Theorem 2.4, which we restate as follows.

Theorem 4.3 *Suppose* $l_1 \geq l_2 \geq \cdots > 0$ *and*

$$l_j \sim Lg(j) \ (= Lh^{-1}(1/j)) \quad as \quad j \to \infty, \tag{4.26}$$

for some constant $L > 0$ *and* $h \in G_d$ *with* $d \in (0,1)$. *Let* $\delta(x) = \sum_{j=1}^{\infty}\{l_j x\}$, *as in* (3.36).
   Then

$$\delta(x) \sim -\zeta(d)L^d f(x), \quad as \quad x \to \infty, \tag{4.27}$$

where $f(x) = 1/h(1/x)$ *and* $\zeta = \zeta(s)$ *denotes the Riemann zeta-function.*

Proof. Let $J(\epsilon)$ be defined by (3.19). Since by hypothesis (4.26),

$$\frac{l_j}{g(j)} = \frac{l_j}{h^{-1}(\frac{1}{j})} \to L \quad \text{as} \quad j \to \infty,$$

it follows from Proposition 3.2(i) that

$$\frac{h(l_j)}{h(h^{-1}(1/j))} = \frac{h(l_j)}{\frac{1}{j}} = jh(l_j) \to L^d, \quad \text{as} \quad j \to \infty.$$

Thus $J(\epsilon)h(l_{J(\epsilon)}) \to L^d$ as $\epsilon \to 0^+$. We know that $\frac{l_{J(\epsilon)}}{\epsilon} \to 1$ as $\epsilon \to 0^+$. (Indeed, since $l_{J(\epsilon)} \geq \epsilon > l_{J(\epsilon)+1}$, $1 \leq l_{J(\epsilon)}/\epsilon < l_{J(\epsilon)}/l_{J(\epsilon)+1} \to 1$, as $\epsilon \to 0^+$, by (4.26) and (3.8).) By Proposition 3.2(i), this implies that $h(l_{J(\epsilon)})/h(\epsilon) \to 1$ as $\epsilon \to 0^+$. Hence

$$J(\epsilon)h(\epsilon) = J(\epsilon)h(l_{J(\epsilon)})\frac{h(\epsilon)}{h(l_{J(\epsilon)})} \to L^d \cdot 1 = L^d, \quad \text{as} \quad \epsilon \to 0^+; \tag{4.28}$$

that is, $J(\epsilon) \sim L^d/h(\epsilon)$ as $\epsilon \to 0^+$. In view of hypothesis (H1), it follows that $J(\epsilon) \to \infty$ as $\epsilon \to 0^+$.
   Let $k \geq 2$ be an arbitrary fixed integer. Using the facts that $J(1/x) \geq J(2/x) \geq \cdots$ and $J(q/x) < j \leq J((q-1)/x)$ implies that $[l_j x] = q - 1$, we

obtain as in [LaPo2, p. 62], upon applying in particular Abel's summation formula, that

$$
\begin{aligned}
\delta(x) &= x \sum_{j>J(1/x)} l_j + \sum_{j\leq J(k/x)} \{l_j x\} + \sum_{q=2}^{k} \sum_{j=J(q/x)+1}^{J(\frac{q-1}{x})} \{l_j x\} \\
&= x \sum_{j>J(1/x)} l_j + \sum_{j\leq J(k/x)} \{l_j x\} + \sum_{q=2}^{k} \sum_{j=J(\frac{q}{x})+1}^{J(\frac{q-1}{x})} (l_j x - (q-1)) \\
&= x \sum_{j>J(\frac{k}{x})} l_j + \sum_{j\leq J(k/x)} \{l_j x\} - \sum_{q=2}^{k} (q-1)(J(\frac{q-1}{x}) - J(\frac{q}{x})) \\
&= x \sum_{j>J(k/x)} l_j + \sum_{j\leq J(k/x)} \{l_j x\} - \sum_{q=1}^{k-1} J(\frac{q}{x}) + (k-1)J(\frac{k}{x}) \\
&=: A + B + C,
\end{aligned}
\tag{4.29}
$$

where

$$
A := x \sum_{j>J(k/x)} l_j, \tag{4.30}
$$

$$
B := kJ(\frac{k}{x}) - \sum_{q=1}^{k-1} J(\frac{q}{x}), \tag{4.31}
$$

$$
C := \sum_{j\leq J(k/x)} (\{l_j x\} - 1). \tag{4.32}
$$

Note that, in (4.29)—(4.32), A, B and C depend on x. We now estimate, for *fixed* k, each of these terms as $x \to \infty$.

We first observe that since $-1 \leq \{l_j x\} - 1 < 0$, we have $-J(k/x) \leq C < 0$; it thus follows from (4.32) and (4.28) that (for fixed k),

$$
\begin{aligned}
0 \leq -L^{-d} h(\frac{1}{x}) C &\leq L^{-d} h(1/x) J(\frac{k}{x}) \\
&= L^{-d} \frac{h(1/x)}{h(k/x)} J(\frac{k}{x}) h(\frac{k}{x}) \\
&\to L^{-d} \cdot k^{-d} \cdot L^d = k^{-d}, \quad \text{as } x \to \infty. \tag{4.33}
\end{aligned}
$$

Next, we deduce from the definition of B and (4.33) that

$$
\begin{aligned}
L^{-d}h(\tfrac{1}{x})B &= L^{-d}h(\tfrac{1}{x})kJ(\tfrac{k}{x}) - L^{-d}h(\tfrac{1}{x})\sum_{q=1}^{k-1} J(\tfrac{q}{x}) \\
&= L^{-d}\frac{h(1/x)}{h(k/x)}kh(\tfrac{k}{x})J(\tfrac{k}{x}) \\
&\quad - L^{-d}\sum_{q=1}^{k-1}\left(\frac{h(1/x)}{h(q/x)}h(\tfrac{q}{x})J(\tfrac{q}{x})\right) \\
&\longrightarrow k^{1-d} - \sum_{q=1}^{k-1} q^{-d}, \quad \text{as } x \to \infty. \qquad (4.34)
\end{aligned}
$$

(This follows since for each fixed $q = 1, \cdots, k$, $h(1/x)/h(q/x) \to q^{-d}$ and by (4.29), $h(q/x)J(q/x) \to L^d$.)

Further, we claim that

$$
L^{-d}h(\tfrac{1}{x})A \to k^{1-d}\frac{d}{1-d}, \quad \text{as } x \to \infty. \qquad (4.35)
$$

Indeed, since $\alpha_j = l_j/g(j) \to L$ as $j \to \infty$ and $J(\epsilon)h(\epsilon) \to L^d$ as $\epsilon \to 0^+$, we see that for each $\epsilon > 0$, there is some $x_0 > 0$ such that for all $x \geq x_0$, we have

$$
\alpha_j \in (L - \epsilon, L + \epsilon), \quad \text{for all } j > J(\tfrac{k}{x}).
$$

Thus, for all $x \geq x_0$,

$$
\begin{aligned}
L^{-d}h(\tfrac{1}{x})A &= L^{-d}xh(\tfrac{1}{x})\sum_{j>J(k/x)} l_j \\
&= L^{-d}xh(\tfrac{1}{x})\sum_{j>J(k/x)} \alpha_j g(j) \\
&\leq L^{-d}(L+\epsilon)xh(\tfrac{1}{x})\int_{J(k/x)}^{\infty} g(t)dt \\
&= L^{-d}(L+\epsilon)xh(\tfrac{1}{x})\frac{\int_{J(k/x)}^{\infty} g(t)dt}{J(\tfrac{k}{x})\, g\left(J(\tfrac{k}{x})\right)}J(\tfrac{k}{x})\, g\left(J(\tfrac{k}{x})\right) \\
&= (L+\epsilon)L^{-d}x\left(\frac{h(1/x)}{h(k/x)}\right)\left(h(\tfrac{k}{x})J(\tfrac{k}{x})\right)
\end{aligned}
$$

$$\cdot \left(\frac{g\left(J(k/x)\right)}{g\left(\frac{1}{h(k/x)}\right)}\right) g\left(\frac{1}{h(k/x)}\right) \frac{\int_{J(k/x)}^{\infty} g(t)dt}{J(k/x)g\left(J(k/x)\right)}$$

$$= (L+\epsilon)L^{-d}x \cdot k^{-d} \cdot L^d \cdot L^{-1} \cdot \frac{k}{x} \cdot \frac{d}{1-d}(1+o(1))$$

$$= (1+\frac{\epsilon}{L})k^{1-d}\frac{d}{1-d}(1+o(1)), \quad \text{as } x \to \infty. \qquad (4.36)$$

This is so since by assumption (H2),

$$\frac{h(1/x)}{h(k/x)} \to k^{-d}, \quad \text{as } x \to \infty,$$

and according to (4.28),

$$h(\frac{k}{x})J(\frac{k}{x}) \to L^d, \quad \text{as } x \to \infty;$$

then, by Proposition 3.2(ii), we obtain

$$\frac{g\left(J(k/x)\right)}{g\left(\frac{1}{h(k/x)}\right)} \to (L^d)^{-1/d} = L^{-1}, \quad \text{as } x \to \infty,$$

and $g(\frac{1}{h(k/x)}) = h^{-1}(h(k/x)) = k/x$. Finally, from Lemma 3.3(1), we see that for $y := J(k/x)$,

$$\frac{\int_y^{\infty} g(t)dt}{yg(y)} \to \frac{d}{1-d}, \quad \text{as } x \to \infty \text{ (which forces } y \to \infty).$$

So putting all these facts together, we deduce (4.36).

Similarly, we can prove that

$$L^{-d}h(\frac{1}{x})A \geq (1-\frac{\epsilon}{L})k^{1-d}\frac{d}{1-d}(1+o(1)), \quad \text{as } x \to \infty. \qquad (4.37)$$

Since ϵ can be arbitrarily small, we see from combining (4.36) and (4.37) that

$$L^{-d}h(\frac{1}{x})A \to k^{1-d}\frac{d}{1-d}, \quad \text{as } x \to \infty. \qquad (4.38)$$

Thus, for *fixed* k, by adding (4.34) and (4.38), we have as $x \to \infty$,

$$L^{-d}h(\frac{1}{x})(A+B) \;\; \to \;\; k^{1-d}\frac{d}{1-d} + k^{1-d} - \sum_{q=1}^{k-1} q^{-d}$$

$$= \;\; \frac{1}{1-d}k^{1-d} - \sum_{q=1}^{k-1} q^{-d}$$

$$= \;\; w_k(d) + \frac{1}{1-d}, \tag{4.39}$$

where

$$w_k(s) := \int_1^k (t^{-s} - [t]^{-s})dt \;\; (= -\frac{1}{1-s} + \frac{1}{1-s}k^{1-s} - \sum_{q=1}^{k-1} q^{-s}).$$

The sequence of entire functions $\{w_k(s)\}_{k=1}^{\infty}$ converges uniformly on every compact subset of $Re\ s > 0$ to the function

$$w(s) := \int_1^{\infty} (t^{-s} - [t]^{-s})dt,$$

which is analytic in the domain $Re\ s > 0$. Further, as noted in (1.4), we have

$$w(s) = -\frac{1}{1-s} - \zeta(s), \quad \text{for} \;\; Re\ s > 0.$$

Hence

$$w_k(d) + \frac{1}{1-d} \to -\zeta(d), \quad \text{as} \;\; k \to \infty. \tag{4.40}$$

Since $L^{-d}h(1/x)\delta(x) = L^{-d}h(1/x)(A+B) + L^{-d}h(1/x)C$, and by (4.33),

$$-k^{-d}(1+o(1)) \;\; \leq L^{-d}h(1/x)C \leq 0,$$

as $x \to \infty$, it follows that

$$L^{-d}h(1/x)(A+B) - k^{-d}(1+o(1)) \;\; \leq \;\; L^{-d}h(1/x)\delta(x)$$
$$\leq \;\; L^{-d}h(1/x)(A+B), \tag{4.41}$$

where '$o(1)$' tends to 0 as $x \to \infty$. From (4.39), we deduce that for each *fixed* $k \geq 2$, if we take both the upper and lower limits in (4.41) as $x \to \infty$, we obtain

$$-k^{-d} + w_k(d) + \frac{1}{1-d} \leq \liminf_{x\to\infty} L^{-d}h(1/x)\delta(x)$$
$$\leq \limsup_{x\to\infty} L^{-d}h(1/x)\delta(x)$$
$$\leq w_k(d) + \frac{1}{1-d}. \qquad (4.42)$$

By letting $k \to \infty$ in (4.42) and using (4.40), we then deduce that

$$\lim_{x\to\infty} L^{-d}h(\frac{1}{x})\delta(x) = -\zeta(d);$$

that is,

$$\delta(x) \sim -L^d\zeta(d)f(x), \quad \text{as } x \to \infty, \qquad (4.43)$$

where $f(x) = 1/h(1/x)$.

Therefore, we have completed the proof of Theorem 4.3, and hence of Theorem 2.4. ∎

Remark 4.4 More precisely, (3.36) and (4.43) clearly imply (2.5) from Theorem 2.4(a). Further, in view of (3.36′), (2.5) applied with $x = \sqrt{\lambda}/\pi$ yields (2.6) from Theorem 2.4(b), since by Lemma 3.1(1),

$$f(\sqrt{\lambda}/\pi) = \pi^{-d}f(\sqrt{\lambda})(1 + o(1)),$$

as $\lambda \to \infty$. This proves Theorem 2.4. Now, Corollary 2.6 follows immediately by combining Theorems 2.4(b) and 2.5.

5   The Complex Zeros of the Riemann Zeta-Function

In the light of Theorem 2.5, we can rephrase Theorem 2.9 as indicated in Theorem 5.1 below. This will extend some of the results in [LaMa1,2] and show, in particular, the influence of complex (or 'critical') zeros of ζ on the spectrum of 'fractal strings'.

Recall that the '*critical*' zeros of ζ are those located in the '*critical strip*' $0 < Re\ s < 1$. They appear in conjugate pairs, symmetrically with respect to the '*critical line*' $Re\ s = 1/2$. Further, the Riemann hypothesis (which is not assumed here) states that they all lie on the 'critical line'. (See, e.g., [Ti], [Od].)

Theorem 5.1 *Suppose $h \in G_d$ is differentiable and $xh'(x)/h(x) \geq \mu > 0$ for all $x > 0$ and for some positive constant μ. Further, Let $\rho = d + i\nu$ ($d \in (0,1), \nu \in \mathbf{R}$) be a zero of the Riemann zeta-function. Then there exists a positive nonincreasing sequence $(l_j)_{j=1}^{\infty}$ which does not satisfy $l_j \sim Lg(j)$ for any constant L (that is, $l_j/g(j)$ does not converge as $j \to \infty$), but such that $\sum_{j=1}^{\infty} l_j < \infty$ and*

$$\mathcal{N}(x) := \sum_{j=1}^{\infty}[l_j x] = (\sum_{j=1}^{\infty} l_j)x + cf(x) + o(f(x)), \quad as \ \ x \to \infty,$$

where c is a nonzero constant. Moreover, with the notation of (3.23), we have $0 < M_*(h;(l_j)) < M^*(h;(l_j)) < \infty$. Here, as before, $g(x) = h^{-1}(1/x)$ and $f(x) = 1/h(1/x)$.

(It will follow from the proof of Theorem 5.1 that $c = \zeta(d) < 0$.)

Without loss of generality, we may assume that $\nu > 0$; indeed, $\zeta(\bar{\rho}) = \overline{\zeta(\rho)} = 0$ and $\nu \neq 0$ since $\zeta(d) \neq 0$ for $d \in (0,1)$.

We now state several definitions and lemmas that will be used in the proof of this theorem.

Definition 5.2 *For a fixed $\beta \in \mathbf{R}$, let*

$$
\begin{aligned}
U(x) &:= f(x) + \beta f(x)x^{i\nu} + \beta f(x)x^{-i\nu} \\
&= f(x)(1 + 2\beta \cos(\nu \ln x)).
\end{aligned} \tag{5.1}
$$

We shall assume from now on that $\beta > 0$ has been chosen so small that the function U is *positive and strictly increasing* on $(0, \infty)$. This can be done since for all $x > 0$,

$$
\begin{aligned}
U'(x) &= f'(x)(1 + 2\beta \cos(\nu \ln x)) + f(x)(-\frac{2\beta\nu}{x}) \sin(\nu \ln x) \\
&\geq f'(x)(1 - 2\beta) + f(x)(-\frac{2\beta\nu}{x}).
\end{aligned}
$$

(Note that since f is positive and increasing, both $f(x)$ and $f'(x)$ are positive.) Thus if

$$
\frac{xf'(x)}{f(x)} > \frac{2\beta\nu}{1 - 2\beta}, \quad \text{then} \quad U'(x) > 0,
$$

provided that $1 - 2\beta > 0$. Further, a simple calculation yields

$$
\frac{xf'(x)}{f(x)} = \frac{yh'(y)}{h(y)},
$$

where $y := 1/x$. Since we assume that $yh'(y)/h(y) \geq \mu > 0$ for all $y > 0$, by choosing β in such a way that $2\beta\nu/(1 - 2\beta) < \mu$, we will deduce that $U'(x) > 0$ for all $x \in (0, \infty)$. Hence U is strictly increasing, provided that $0 < \beta < \mu/2(\mu + \nu)$.

This enables us to define the positive strictly increasing sequence $(m_j)_{j=1}^{\infty}$ by

$$
U(m_j) = j \quad (j = 1, 2, \ \ldots), \tag{5.2}
$$

and then the strictly decreasing sequence $(l_j)_{j=1}^{\infty}$ by

$$
l_j = m_j^{-1} \quad (j = 1, 2, \ldots). \tag{5.3}
$$

We next let

$$
T(x) := [U(x)] \quad (x > 0). \tag{5.4}
$$

An immediate consequence of these definitions is that

$$
T(x) = \#\{j \geq 1 : m_j \leq x\} = \#\{j \geq 1 : l_j \geq 1/x\}. \tag{5.5}
$$

So we can say that T is the 'counting function' associated with the sequence $(l_j)_{j=1}^{\infty}$.

Note that $\sum_{j=1}^{\infty} l_j < \infty$. Indeed,

$$h(1/x)T(x) \leq h(1/x)U(x) = 1 + 2\beta \cos(\nu \ln x) \leq 1 + 2\beta, \quad \text{for all } x > 0.$$

In particular, for $x = m_j$, we obtain $h(l_j)j \leq 1 + 2\beta$; thus, by hypothesis (H3) and Proposition 3.2(ii),

$$l_j \leq h^{-1}(\frac{1+2\beta}{j}) = g(\frac{1}{1+2\beta}j) \sim (1+2\beta)^{1/d}g(j), \quad \text{as } j \to \infty.$$

So $l_j \leq C_1 g(j)$ for all j large enough, where C_1 is a positive constant. Since for all $j \geq 1$, $\sum_{k=j+1}^{\infty} g(k) < \int_j^{\infty} g(x)dx < \infty$ as was noted before, we see that $\sum_{j=1}^{\infty} l_j < \infty$.

Similarly, $h(1/x)T(x) \geq 1 - 2\beta > 0$ and so $l_j \geq C_2 g(j)$ for some positive constant C_2. Hence, $l_j \asymp g(j)$ as $j \to \infty$ and by Theorem 2.7 (or 3.4), it follows that $0 < M_*(h; (l_j)) \leq M^*(h; (l_j)) < \infty$.

We will now verify that the sequence $(l_j)_{j=1}^{\infty}$ satisfies the claim of the theorem.

Definition 5.3 [LaMa2, Definition 3.18] *Let $V = V(u)$ be a continuous complex-valued function for $u \geq 1$ and such that $|V(u)| = O(u^{\eta})$ as $u \to \infty$, for some constant $\eta \in (0,1)$. Let k be a positive integer ≥ 2. We then define:*

$$
\begin{aligned}
B(x,k,V) &:= x\left([u^{-1}V(u)]_{u=x/k}^{u=x} + \int_{x/k}^{x} V(u)u^{-2}du\right) \\
&\quad - \sum_{p=1}^{k-1} p\left(V(\frac{x}{p}) - V(\frac{x}{p+1})\right) - V(x) + x\int_{x}^{\infty} V(u)u^{-2}du \\
&= -kV(\frac{x}{k}) + x\int_{x/k}^{\infty} V(u)u^{-2}du \\
&\quad - \sum_{p=1}^{k-1} p\left(V(\frac{x}{p}) - V(\frac{x}{p+1})\right).
\end{aligned}
\tag{5.6}
$$

We shall see later on that the evaluation of $\delta(x)$ $(= \sum_{j=1}^{\infty}\{l_j x\})$ leads to a linear combination of terms of the form $B(x,k,V)$. (Cf. Eqs. (5.16) and (5.17) below.)

The following simple lemma [LaMa2, Lemma 3.19] supplements Eq. (1.4).

Lemma 5.4 *Let s with $0 < Re\ s < 1$ and let $k \geq 2$. Then we have*

$$\zeta(s) = \frac{1}{s-1} + \int_1^\infty ([u]^{-s} - u^{-s})du$$

$$= \frac{1}{s-1} + \int_1^k ([u]^{-s} - u^{-s})du + O(k^{-Re\ s}). \qquad (5.7)$$

More precisely, there exists a positive constant W (independent of k and depending only on s) such that

$$\left|\zeta(s) - \left(\frac{1}{s-1} + \int_1^k ([u]^{-s} - u^{-s})du\right)\right| \leq Wk^{-Re\ s},$$

for all $k \geq 2$. (We may choose, e.g., $W = |s|/Re\ s$.)

Now we are ready to establish the following key lemma. It is not quite as sharp as [LaMa2, Lemma 3.20] due to the additional complication caused by the gauge function h. (See Remark 5.6(a).) It will enable us to extend some of our results in Section 4.2 obtained for real numbers d in the 'critical interval' $(0,1)$ to complex numbers $s = d + i\nu$ in the 'critical strip' $0 < Re\ s < 1$. (See Remark 5.6(b) and Example 7.3.)

Lemma 5.5 *Suppose $h \in G_d$ for some $d \in (0,1)$. Let $V(x) = f(x)x^{i\nu}$ with $\nu \in \mathbf{R}$. Then we have for every fixed integer k,*

$$B(x,k,V) = -\zeta(d+i\nu)f(x)x^{i\nu} + O(k^{-d}f(x)), \quad as\ x \to \infty. \qquad (5.8)$$

More precisely, there exists a positive constant Q (independent of $k \geq 2$ or $x > 0$) with the following property: for any fixed integer $k \geq 2$, there is $x_0 = x_0(k) > 0$ such that for all $x > x_0(k)$,

$$|B(x,k,V) + \zeta(d+i\nu)f(x)x^{i\nu}| \leq Qk^{-d}f(x).$$

Proof. Let

$$C(x,k,V) = -k^{1-d-i\nu}f(x)x^{i\nu} + \frac{k^{1-(d+i\nu)}}{1-(d+i\nu)}f(x)x^{i\nu}$$

$$- \sum_{p=1}^{k-1}(p^{1-(d+i\nu)} - p(p+1)^{-d-i\nu})f(x)x^{i\nu}$$

$$= f(x)x^{i\nu}\left(-k^{1-(d+i\nu)} + \frac{k^{1-(d+i\nu)}}{1-(d+i\nu)}\right.$$

$$\left. - \sum_{p=1}^{k-1}\left(p^{1-(d+i\nu)} - p(p+1)^{-(d+i\nu)}\right)\right).$$

First, we want to show that for each fixed $k \geq 2$,

$$\lim_{x \to \infty} \frac{B(x,k,V) - C(x,k,V)}{f(x)} = 0. \tag{5.9}$$

Since $V(x) = f(x)x^{i\nu}$, it follows from (5.6) that

$$\begin{aligned}
B(x,k,V) &= -kf(x/k)x^{i\nu}k^{-i\nu} + x\int_{x/k}^{\infty} f(u)u^{-2+i\nu}du \\
&\quad - \sum_{p=1}^{k-1} p\left(x^{i\nu}p^{-i\nu}f(\frac{x}{p}) - x^{i\nu}(p+1)^{-i\nu}f(\frac{x}{p+1})\right).
\end{aligned}$$

By Lemma 3.1(1), for fixed $k \geq 2$, $f(x/p) - p^{-d}f(x) = o(f(x))$ uniformly for $p = 1, 2, \ldots, k$ as $x \to \infty$. Thus it is clear that

$$\begin{aligned}
&\sum_{p=1}^{k-1} p\left(x^{i\nu}p^{-i\nu}f(\frac{x}{p}) - x^{i\nu}(p+1)^{-i\nu}f(\frac{x}{p+1})\right) \\
&\quad - \left(\sum_{p=1}^{k-1}(p^{1-(d+i\nu)} - p(p+1)^{-d-i\nu})f(x)x^{i\nu}\right) \\
&= \sum_{p=1}^{k-1} p\left(x^{i\nu}p^{-i\nu}(f(\frac{x}{p}) - p^{-d}f(x))\right) \\
&\quad - \sum_{p=1}^{k-1} p\left(x^{i\nu}(p+1)^{-i\nu}(f(\frac{x}{p+1}) - (p+1)^{-d}f(x))\right) \\
&= o(f(x)), \tag{5.10}
\end{aligned}$$

as $x \to \infty$.

Moreover, by Lemma 3.3(2) applied to $z = 1 - i\nu$, we see that

$$\frac{\int_y^{\infty} f(u)u^{-2+i\nu}du}{y^{-1+i\nu}f(y)} \to \frac{1}{1-(d+i\nu)}, \quad \text{as } y \to \infty.$$

If we take $y = x/k$, we have as $x \to \infty$:

$$\frac{\int_{x/k}^{\infty} f(u)u^{-2+i\nu}du}{(\frac{x}{k})^{-1+i\nu}f(x/k)} \to \frac{1}{1-(d+i\nu)};$$

that is,

$$\frac{x \int_{x/k}^{\infty} f(u)u^{-2+i\nu}du}{x^{i\nu}f(x/k)} \rightarrow \frac{k^{1-i\nu}}{1-(d+i\nu)}.$$

It then follows from Lemma 3.1(1) that

$$\frac{x \int_{x/k}^{\infty} f(u)u^{-2+i\nu}du}{x^{i\nu}f(x)} = \frac{x \int_{x/k}^{\infty} f(u)u^{-2+i\nu}du}{x^{i\nu}f(x/k)} \frac{f(x/k)}{f(x)}$$

$$\rightarrow \frac{k^{1-i\nu}}{1-(d+i\nu)}k^{-d} = \frac{k^{1-(d+i\nu)}}{1-(d+i\nu)}, \quad \text{as } x \rightarrow \infty;$$

thus, as $x \rightarrow \infty$,

$$x \int_{x/k}^{\infty} f(u)u^{-2+i\nu}du - \frac{k^{1-(d+i\nu)}}{1-(d+i\nu)}x^{i\nu}f(x) = o(f(x)). \tag{5.11}$$

Finally,

$$-kf(\frac{x}{k})x^{i\nu}k^{-i\nu} + k^{1-d-i\nu}f(x)x^{i\nu}$$

$$= -kx^{i\nu}k^{-i\nu}\left(f(\frac{x}{k}) - k^{-d}f(x)\right) = o(f(x)), \tag{5.12}$$

as $x \rightarrow \infty$.

Now if we put (5.10)—(5.12) together, we can easily see that (5.9) holds. Next, for fixed k, we consider the behavior of $C(x,k,V)$ as $x \rightarrow \infty$.

$$C(x,k,V) = f(x)x^{i\nu}\left(-k^{1-(d+i\nu)} + \frac{k^{1-(d+i\nu)}}{1-(d+i\nu)}\right.$$

$$\left. -\sum_{p=1}^{k-1}(p^{1-(d+i\nu)} - p(p+1)^{-(d+i\nu)})\right)$$

$$= f(x)x^{i\nu}\left(-k^{1-(d+i\nu)} + \frac{k^{1-(d+i\nu)}}{1-(d+i\nu)} - (1-k^{1-(d+i\nu)})\right.$$

$$\left. -\sum_{p=1}^{k-1}(p+1)^{-(d+i\nu)}\right)$$

$$= f(x)x^{i\nu}\left(-\sum_{p=0}^{k-1}(p+1)^{-(d+i\nu)} + \frac{k^{1-(d+i\nu)}}{1-(d+i\nu)}\right)$$

$$
\begin{aligned}
&= \quad f(x)x^{i\nu}\left(-\left(\sum_{p=0}^{k-2}(p+1)^{-(d+i\nu)} - \frac{k^{1-(d+i\nu)}}{1-(d+i\nu)}\right.\right. \\
&\qquad \left.\left. + \frac{1}{1-(d+i\nu)}\right) + \frac{1}{1-(d+i\nu)} - k^{-(d+i\nu)}\right) \\
&= \quad f(x)x^{i\nu}\left(-\int_1^k ([u]^{-(d+i\nu)} - u^{-(d+i\nu)})du \right. \\
&\qquad \left. + \frac{1}{1-(d+i\nu)} - k^{-(d+i\nu)}\right) \\
&= \quad f(x)x^{i\nu} \cdot A, \quad \text{say.} \tag{5.13}
\end{aligned}
$$

(To derive the second equality in (5.13), we just wrote $p = (p+1) - 1$.)

By (5.7), there exists a positive constant Q_1 (independent of k) such that for all $k \geq 2$,

$$
|\zeta(d+i\nu) + A| \leq Q_1 k^{-d}.
$$

(With the notation of Lemma 5.4, we may choose $Q_1 = W+1 = |d+i\nu|/d+1$.) Thus, for fixed k, we obtain that

$$
\begin{aligned}
&|C(x,k,V) + f(x)x^{i\nu}\zeta(d+i\nu)| \\
&= \quad f(x)|\zeta(d+i\nu) + A| \leq Q_1 k^{-d}f(x), \quad \text{for all} \quad x > 0. \tag{5.14}
\end{aligned}
$$

So if we put (5.9) and (5.14) together, we complete the proof of (5.8), for any fixed k. More precisely, we deduce from (5.9) that for any fixed $k \geq 2$, there exists $x_0(k) > 0$ such that for all $x > x_0(k)$,

$$
|B(x,k,V) - C(x,k,V)| \leq k^{-d}f(x).
$$

Hence, by (5.14), we have for all $x > x_0(k)$:

$$
\begin{aligned}
&|B(x,k,V) + f(x)x^{i\nu}\zeta(d+i\nu)| \\
&\leq \quad |C(x,k,V) + f(x)x^{i\nu}\zeta(d+i\nu)| + k^{-d}f(x) \\
&\leq \quad (Q_1+1)k^{-d}f(x) =: Qk^{-d}f(x),
\end{aligned}
$$

with $Q = W + 2 = |d + i\nu|/d + 2$, as desired. This proves Lemma 5.5. ∎

We are now ready to prove Theorem 5.1.

Proof of Theorem 5.1. Let $(l_j)_{j=1}^\infty$ be the sequence defined as above. Let $\delta(x)$ be defined by (3.36). Since $\sum_{j=1}^\infty l_j < \infty$, we have

$$\mathcal{N}(x) := \sum_{j=1}^\infty [l_j x] = (\sum_{j=1}^\infty l_j)x - \sum_{j=1}^\infty \{l_j x\} = (\sum_{j=1}^\infty l_j)x - \delta(x). \qquad (5.15)$$

In the sequel, we will show that for any fixed integer $k \geq 2$,

$$\delta(x) = \delta_1(x) + \delta_2(x) + \delta_3(x) + O(k^2 + k^{-d}f(x)), \text{ as } x \to \infty, \qquad (5.16)$$

where

$$\begin{aligned}
\delta_1(x) &= B(x, k, V_1) \text{ with } V_1(x) = f(x); \\
\delta_2(x) &= \beta B(x, k, V_2) \text{ with } V_2(x) = f(x)x^{i\nu}; \\
\delta_3(x) &= \beta B(x, k, V_3) \text{ with } V_3(x) = f(x)x^{-i\nu}.
\end{aligned} \qquad (5.17)$$

More precisely, (5.16) means that for any given integer $k \geq 2$, there exist positive constants E and $x_0(k)$ such that for all $x > x_0(k)$,

$$|\delta(x) - (\delta_1(x) + \delta_2(x) + \delta_3(x))| \leq E(k^2 + k^{-d}f(x)).$$

Assuming this for now, then we deduce from Lemma 5.5 that, for any fixed k, we have as $x \to \infty$,

$$\begin{aligned}
B(x, k, V_1) &= -\zeta(d)f(x) + O(k^{-d}f(x)); \\
B(x, k, V_2) &= -\zeta(d + i\nu)f(x) + O(k^{-d}f(x)); \\
B(x, k, V_3) &= -\zeta(d - i\nu)f(x) + O(k^{-d}f(x)).
\end{aligned} \qquad (5.18)$$

Thus, from (5.16), we see that the key point is that for any given $\epsilon > 0$, we need to choose some positive integer $k_0(\epsilon)$ such that

$$k_0(\epsilon)^2 + k_0(\epsilon)^{-d}f(x) < \epsilon f(x), \text{ for all } x \text{ large enough}. \qquad (5.19)$$

For any given $\epsilon > 0$, let us choose k_ϵ in such a way that for all $k \geq k_\epsilon$, we have $k^{-d} < \epsilon/2$. For example, we can choose $k_\epsilon = [(\epsilon/2)^{-1/d}] + 1$. Then for this fixed k_ϵ, there exists $x_1(k_\epsilon) > 0$, such that for all $x > x_1(k_\epsilon)$, we have $k_\epsilon^2 < \frac{\epsilon}{2}f(x)$ (since f is increasing, and by (H1), $f(x) \to \infty$ as $x \to \infty$). Thus

$$k_\epsilon^2 + k_\epsilon^{-d}f(x) < \frac{\epsilon}{2}f(x) + \frac{\epsilon}{2}f(x) = \epsilon f(x), \text{ for all } x > x_1(k_\epsilon). \qquad (5.20)$$

We claim that for this fixed k_ϵ, we have

$$\begin{aligned}
\delta(x) &= \delta_1(x) + \delta_2(x) + \delta_3(x) + O(k_\epsilon^2 + k_\epsilon^{-d} f(x)) \\
&= -\zeta(d) f(x) - \beta\zeta(d + i\nu) f(x) x^{i\nu} \\
&\quad - \beta\zeta(d - i\nu) f(x) x^{-i\nu} + o(f(x)).
\end{aligned} \tag{5.21}$$

This is so because for some constant $E > 0$ and all $x > x_1(k_\epsilon)$,

$$|\delta(x) - (\delta_1(x) + \delta_2(x) + \delta_3(x))| \le E(k_\epsilon^2 + k_\epsilon^{-d} f(x)) < E\epsilon f(x). \tag{5.22}$$

Also, by (5.18), there exist positive numbers $x_2(k_\epsilon), x_3(k_\epsilon)$ and $x_4(k_\epsilon)$ such that

$$\begin{aligned}
|\delta_1(x) + \zeta(d) f(x)| &= |B(x, k_\epsilon, V_1) + \zeta(d) f(x)| \\
&\le Q k_\epsilon^{-d} f(x) < Q\epsilon f(x),
\end{aligned} \tag{5.23}$$

for all $x > x_2(k_\epsilon)$;

$$\begin{aligned}
|\delta_2(x) + \beta\zeta(d + i\nu) f(x) x^{i\nu}| &= \beta |B(x, k_\epsilon, V_2) + \zeta(d + i\nu) f(x) x^{i\nu}| \\
&\le \beta Q k_\epsilon^{-d} f(x) < \beta Q\epsilon f(x),
\end{aligned} \tag{5.24}$$

for all $x > x_3(k_\epsilon)$;

$$\begin{aligned}
|\delta_3(x) + \zeta(d - i\nu) f(x) x^{-i\nu}| &= \beta |B(x, k, V_3) + \zeta(d - i\nu) f(x) x^{-i\nu}| \\
&\le \beta Q k_\epsilon^{-d} f(x) < \beta Q\epsilon f(x),
\end{aligned} \tag{5.25}$$

for all $x > x_4(k_\epsilon)$. Now let $x_0(\epsilon) = x_0(k_\epsilon) = \max(x_1(k_\epsilon), x_2(k_\epsilon), x_3(k_\epsilon), x_4(k_\epsilon))$. Then, for all $x > x_0(\epsilon)$, combining (5.22)—(5.25) together, we have

$$\begin{aligned}
&|\delta(x) + \zeta(d) f(x) + \beta\zeta(d + i\nu) f(x) x^{i\nu} + \beta\zeta(d - i\nu) f(x) x^{-i\nu}| \\
&< (E + 2Q\beta)\epsilon f(x),
\end{aligned}$$

which proves (5.21). We note that k_ϵ acts like a 'bridge' between the above estimates.

So if we assume, as above, that $\zeta(d + i\nu) = \zeta(d - i\nu) = 0$, we see from (5.21) that

$$\delta(x) = -\zeta(d) f(x) + o(f(x)), \quad \text{as } x \to \infty. \tag{5.26}$$

Thus we deduce from (5.15) and (5.26) that

$$\mathcal{N}(x) = (\sum_{j=1}^{\infty} l_j)x + \zeta(d)f(x) + o(f(x)), \quad \text{as } x \to \infty.$$

Next, we claim that $l_j/g(j)$ does not converge as $j \to \infty$. Indeed, suppose it does converge to some $L \in (0, \infty)$. Since

$$U(m_j) = f(m_j)(1 + 2\beta \cos(\nu \ln m_j)) = j,$$

we then have

$$h(l_j)j = h(1/m_j)j = h(1/m_j)U(m_j) = 1 + 2\beta \cos(\nu \ln m_j).$$

Therefore, $(h(l_j)j)_{j=1}^{\infty}$ does not converge. But if

$$\frac{l_j}{g(j)} = \frac{l_j}{h^{-1}(1/j)} \to L, \quad \text{as } j \to \infty,$$

it follows from Proposition 3.2(i) that

$$\frac{h(l_j)}{h(h^{-1}(1/j))} = h(l_j)j \to L^d, \quad \text{as } j \to \infty.$$

Hence we obtain a contradiction. So $(l_j/g(j))_{j=1}^{\infty}$ does not converge.

Thus we have verified that the sequence $(l_j)_{j=1}^{\infty}$ satisfies all the required properties.

Now all we have to show to conclude the proof of the theorem is that (5.16) holds.

Let $k \geq 2$ be an arbitrary fixed integer. Then

$$\begin{aligned}
\delta(x) &= \sum_{j=1}^{\infty} \{l_j x\} = \sum_{j=1}^{\infty} \{\frac{x}{m_j}\} \\
&= \sum_{m_j > x} x m_j^{-1} + \sum_{p=1}^{k-1} \sum_{\frac{x}{p+1} < m_j < \frac{x}{p}} (x m_j^{-1} - p) + \sum_{m_j \leq \frac{x}{k}} \{m_j^{-1} x\}. \quad (5.27)
\end{aligned}$$

The last equality follows from the fact that if $\frac{x}{p+1} < m_j \leq \frac{x}{p}$, then $\{m_j^{-1}x\} = x m_j^{-1} - [x m_j^{-1}] = x m_j^{-1} - p$. Also note that $T(u) = U(u) + O(1)$. We write:

$$\sum_{p=1}^{k-1} \sum_{\frac{x}{p+1} < m_j \leq \frac{x}{p}} m_j^{-1} = \sum_{\frac{x}{k} < m_j \leq x} m_j^{-1} = \int_{x/k}^{x} u^{-1} dT(u)$$

$$= [u^{-1}T(u)]_{u=x/k}^{u=x} + \int_{x/k}^{x} T(u)u^{-2}du$$

$$= [u^{-1}U(u)]_{u=x/k}^{u=x} + \int_{x/k}^{x} U(u)u^{-2}du + O(kx^{-1}).$$

$$(5.28)$$

Further, we obtain

$$\sum_{p=1}^{k-1} \sum_{\frac{x}{p-1}<m_j\le\frac{x}{p}} p = \sum_{p=1}^{k-1} p\left(T(\frac{x}{p}) - T(\frac{x}{p+1})\right)$$

$$= \sum_{p=1}^{k-1} p\left(U(\frac{x}{p}) - U(\frac{x}{p+1})\right) + O(k^2) \qquad (5.29)$$

and

$$\sum_{m_j>x} m_j^{-1} = \int_x^\infty u^{-1}dT(u)$$

$$= [u^{-1}T(u)]_{u=x}^\infty + \int_x^\infty T(u)u^{-2}du$$

$$= x^{-1}U(x) + \int_x^\infty U(u)u^{-2}du + O(x^{-1}). \qquad (5.30)$$

Finally, by Lemma 3.1(1) (applied for a *fixed k*),

$$\sum_{m_j\le\frac{x}{k}} \{m_j^{-1}x\} \le \sum_{m_j\le\frac{x}{k}} 1 = T(\frac{x}{k})$$

$$= O(f(\frac{x}{k})) = O(k^{-d}f(x)), \quad \text{as } x\to\infty. \qquad (5.31)$$

By putting (5.28)—(5.31) into (5.27), we then obtain:

$$\delta(x) = B(x,k,U) + O(k^2 + k^{-d}f(x)), \quad \text{as } x\to\infty,$$

where $U(x) = f(x) + \beta f(x)x^{i\nu} + \beta f(x)x^{-i\nu}$; i.e., $U = V_1 + \beta V_2 + \beta V_3$. Since obviously $B(x,k,U)$ is linear in U, we deduce that (5.16) holds, as desired. This finally completes the proof of our theorem. ∎

Remark 5.6 (*a*) The result we established above is slightly different from the corresponding one in [LaMa2]. In the present case of general gauge functions, we are not able to obtain the sharper estimate in [LaMa] which says that $\delta(x) = B(x, k, U) + O(k^2 + k^{-d}x^d)$, for all x *and* k large enough. (Compare [LaMa2, Lemma 3.20] with Lemma 5.5 above.) This explains the difference between [LaMa2, Theorem 2.4] and our formulation of Theorem 2.4.

(b)   The above construction remains valid even if $d + i\nu$ (and hence $d - i\nu$) is *not* assumed to be a zero of ζ. Of course, it then leads to a different conclusion, as will be seen in Example 7.3 below.

(c) The work in [LaPo1,2] and then [LaMa1,2] raised the question of finding a suitable notion of 'complex dimension'; see, in particular, [LaPo2, §4.4.*b*], [LaMa2, Remark 3.21(d)], as well as [La3, Fig. 3.1, p. 165] and [La4]. (Indeed, this was the main heuristic motivation leading to the counterexample in [LaMa2], revisited here.) Our present work may shed some light on this question and suggest how to view it in a broader context.

6 Error Estimates for $n \geq 2$

In this section, we provide the proof of Theorem 2.12. Hence, $\Omega \subset \mathbf{R}^n$ is a nonempty open set with finite volume: $|\Omega|_n < \infty$. Further, we assume as in Theorem 2.12 that its boundary $\Gamma = \partial\Omega$ has finite upper h-Minkowski content for some gauge function h: $M^* = M^*(h; \Gamma) < \infty$.

We should keep in mind that in this section, we assume that h satisfies hypothesis (C1)—(C4). Also, we will only consider the Dirichlet Laplacian, even though the result can be easily generalized as in [La1] to higher order operators and Neumann boundary conditions. (See Remark 6.11(b) below.) We found that we can still use many notations, lemmas and propositions in [La1] or [Mt] without any change. We recall some of them, for completeness. (For more details, see [La1, §4.1] and the references therein.)

Definition 6.1 *Let $(X, \| \cdot \|_X)$ be a normed linear space and $B \subset X$. Given any integer $j \geq 0$, the j-width of B in X is given by*

$$d_j(B; X) = \inf_{X_j} \sup_{x \in B} \inf_{y \in X_j} \| x - y \|_X,$$

where the left-most infimum is taken over all j-dimensional subspaces X_j of X.

Definition 6.2 *Let (W, H, b) be a 'variational triple'; that is, W is a dense subspace of the Hilbert space H and b is a bounded, hermitian and coercive form on W. Let $S_b = S_b(W) = \{u \in W : b(u, u) \leq 1\}$. Then we define*

$$N(\lambda; W, H, b) = \#\{j : d_j(S_b(W); H) \geq \lambda^{-1/2}\}. \tag{6.1}$$

It can actually be shown that

$$N(\lambda; W, H, b) = \#\{j \geq 1 : \mu_j \leq \lambda\},$$

69

where $0 < \mu_1 \leq \mu_2 \leq \cdots$ are the eigenvalues of the bounded positive self-adjoint operator T (also assumed to be compact) on the Hilbert space $(W, b(\cdot, \cdot))$, defined by $b(Tu, v) = (u, v)_H$, for all $u, v \in H$.

For any $\omega \subset \Omega$, we let

$$a_\omega(u, v) := \int_\omega \sum_{k=1}^n \frac{\partial u}{\partial x_k} \frac{\partial \bar{v}}{\partial x_k} dx. \qquad (6.2)$$

Thus, using this notation, the 'eigenvalue counting function' $N(\lambda)$ of our problem (P) can be written as

$$N(\lambda) = N(\lambda; H_0^1(\Omega), L^2(\Omega), a_\Omega). \qquad (6.3)$$

Proposition 6.3 [Mt, Proposition 2.7] *Let (W, H, b) be a variational triple and let W_0 be a closed subspace of W. For $\lambda > 0$, let $Z_\lambda := \{u \in W : b(u, v) = \lambda(u, v)_H$, for all $v \in W_0\}$. Then*

$$N(\lambda; W, H, b) = N(\lambda; W_0, H, b) + N(\lambda; Z_\lambda, H, b) - dim(W_0 \cap Z_\lambda).$$

In particular,

$$N(\lambda; W_0, H, b) \leq N(\lambda; W, H, b) \leq N(\lambda; W_0, H, b) + N(\lambda; Z_\lambda, H, b). \qquad (6.4)$$

Besides the classical Sobolev spaces $H_0^1(\Omega)$ and $H^1(\Omega)$, we will also use the following (generalized) Sobolev space:

$$\mathcal{H}_0^1(\Omega) := \{u \in H_0^1(\mathbf{R}^n) : D^\alpha u(x) = 0 \text{ for } |\cdot| - a.e. \ x \in \mathbf{R}^n \setminus \Omega$$
$$\text{and for all } |\alpha| \leq m\},$$

where

$$\alpha = (\alpha_1, \alpha_2, \ldots, \alpha_n) \in \mathbf{N}^n, \quad |\alpha| = \alpha_1 + \alpha_2 + \cdots + \alpha_n$$

and with $x = (x_1, x_2, \cdots, x_n) \in \mathbf{R}^n$,

$$D^\alpha = \frac{\partial^{|\alpha|}}{\partial x_1^{\alpha_1} \partial x_2^{\alpha_2} \cdots \partial x_n^{\alpha_n}}.$$

Proposition 6.4 *There exists a constant $c > 0$ such that for all $\epsilon > 0$, all open (n-dimensional) cubes $Q \subset \Omega$ of side ϵ and all $\lambda, \tau > 0$, we have:*
(i)

$$|N(\lambda; W_Q, L^2(Q), a_Q) - (2\pi)^{-n} \mathcal{B}_n \epsilon^n \lambda^{n/2}| \leq c(1 + \epsilon^{n-1} \lambda^{\frac{n-1}{2}}), \qquad (6.5)$$

where W_Q denotes any one of the spaces $H_0^1(\Omega), \mathcal{H}_0^1(\Omega)$ or $H^1(\Omega)$.
(ii)

$$N(\tau; Z_\lambda(Q), L^2(Q), a_Q) \leq c(1 + \epsilon^{n-1}(\lambda^{\frac{n-1}{2}} + \tau^{\frac{n-1}{2}})), \qquad (6.6)$$

where $Z_\lambda(Q) := \{u \in H^1(\Omega) : a_Q(u, v) = \lambda(u, v)_{L^2(Q)}, \text{ for all } v \in H_0^1(\Omega)\}$, by analogy with Proposition 6.3.

Definition 6.5 *Let the Hilbert space W be continuously embedded in $L^2(\Omega)$ and let ω be an open subset of Ω. Let $S(W)$ be the closed unit ball of W and $S(W)_{/\omega}$ the set of restrictions to ω of elements of $S(W)$. Let $d_j(S(W)_{/\omega}; L^2(\omega))$ be the j-width of $S(W)_{/\omega}$ in $L^2(\omega)$. Then for $\lambda > 0$, set*

$$N^*(\lambda; W, L^2(\omega)) = \#\{j \geq 0 : d_j(S(W)_{/\omega}; L^2(\omega)) \geq \lambda^{-1/2}\}. \qquad (6.7)$$

The next result [Mt] will enable us to obtain boundary estimates for the Dirichlet problem.

Proposition 6.6 *Let Ω be an arbitrary open set in $\mathbf{R}^n$ with finite volume and let ω be an open subset of Ω. Then there exists a positive constant C— depending only on n—such that for all $\lambda > 0$,*

$$N^*(\lambda; \mathcal{H}_0^1(\Omega), L^2(\omega)) \leq C|\omega_{\sqrt{n}\lambda^{-1/2}} \cap \Omega|\lambda^{\frac{n}{2}}, \qquad (6.8)$$

where $\omega_\epsilon = \{x \in \mathbf{R}^n : d(x, \omega) < \epsilon\}$.

We will also adopt the following notation used in [La1].
For $\lambda > 0$ and ω an open subset of Ω, let

$$N_0(\lambda; \omega) = N(\lambda; H_0^1(\omega), L^2(\Omega), a_\Omega)$$

and (by analogy with (6.7))

$$\begin{aligned} N_0^*(\lambda; \omega) &= N^*(\lambda; H_0^1(\Omega), L^2(\omega)), \\ N_1^*(\lambda; \omega) &= N^*(\lambda; \mathcal{H}_0^1(\Omega), L^2(\omega)). \end{aligned} \qquad (6.9)$$

Further, we set

$$\varphi(\lambda; \omega) = (2\pi)^{-n} \mathcal{B}_n |\omega|_n \quad \text{and} \quad \varphi(\lambda) = \varphi(\lambda; \Omega).$$

With this notation, what we want to prove is that

$$|N(\lambda) - \varphi(\lambda)| = O(f(\sqrt{\lambda})), \quad \text{as} \quad \lambda \to \infty,$$

where $f(x) = 1/h(1/x)$, as before.

We now proceed as in [La1, §4.2], but also take into account the fact that we work with general gauge functions.

Proposition 6.7 (estimate near the boundary) *There exist positive constants ϵ_0, λ_0 and c_0 such that for all $0 < \epsilon < \epsilon_0$ and all $\lambda \geq \lambda_0 \epsilon^{-2}$, we have*

$$N_1^*(\lambda; \tilde{\Gamma}_\epsilon) \leq c_0 \epsilon^n \frac{1}{h(\epsilon)} \lambda^{\frac{n}{2}}, \tag{6.10}$$

where $\tilde{\Gamma}_\epsilon := \Gamma_\epsilon \cap \Omega$.

Proof. Fix $\epsilon > 0$; then by (6.8) with $\omega = \tilde{\Gamma}_\epsilon$, we deduce that there exists $C > 0$ such that

$$N_1^*(\lambda; \tilde{\Gamma}_\epsilon) \leq C |(\tilde{\Gamma}_\epsilon)_\delta \cap \Omega| \lambda^{\frac{n}{2}}, \quad \text{for all} \quad \lambda > 0, \tag{6.11}$$

with $\delta = \sqrt{n}\lambda^{-1/2}$. Note that $(\tilde{\Gamma}_\epsilon)_\delta \cap \Omega \subset \tilde{\Gamma}_{\epsilon+\delta}$.

Since we assumed that $M^*(h; \Gamma) := \limsup_{\epsilon \to 0^+} \epsilon^{-n} h(\epsilon) |\tilde{\Gamma}_\epsilon|_n < \infty$, we see that there exist positive constants ϵ_0' and c such that

$$|\Gamma_\epsilon \cap \Omega|_n \leq c \epsilon^n \frac{1}{h(\epsilon)}, \quad \text{for all} \quad \epsilon < \epsilon_0'. \tag{6.12}$$

Now if we take any $\lambda_0 > 0$ and $\epsilon_0 = \epsilon_0'(1 + \sqrt{n}\lambda_0^{-1/2})^{-1}$, then for $\epsilon < \epsilon_0$ and $\lambda \geq \lambda_0 \epsilon^{-2}$, we have $\epsilon + \delta = \epsilon + \sqrt{n}\lambda^{-1/2} \leq (1 + \sqrt{n}\lambda_0^{-1/2})\epsilon < \epsilon_0'$. Thus, by (6.11),

$$\begin{aligned}
N_1^*(\lambda; \Gamma_\epsilon \cap \Omega) &\leq C |\Gamma_{\epsilon+\delta} \cap \Omega| \lambda^{\frac{n}{2}} \leq Cc \frac{(\epsilon + \delta)^n}{h(\epsilon + \delta)} \lambda^{\frac{n}{2}} \\
&\leq c_0 \epsilon^n \frac{1}{h(\epsilon)} \lambda^{\frac{n}{2}}, \quad \text{for all } \epsilon < \epsilon_0 \text{ and } \lambda \geq \lambda_0 \epsilon^{-2},
\end{aligned}$$

where $c_0 > 0$ is a suitable constant.

Note that since $\delta = \sqrt{n}\lambda^{-1/2} \leq \sqrt{n}\epsilon/\sqrt{\lambda_0}$ for all $\lambda \geq \lambda_0\epsilon^{-2}$, we have $(\epsilon + \delta)^n \leq k\epsilon^n$ for some constant $k > 0$. By (C1), h is nondecreasing, and so $h(\epsilon + \delta) \geq h(\epsilon)$. Thus we have $(\epsilon + \delta)^n/h(\epsilon + \delta) \leq k\epsilon^n/h(\epsilon)$, which justifies the last inequality above. ∎

Before we state and prove the next proposition, we need to recall the following construction. We construct as in [La1, §4.2] a sequence of tessellations $\{Q_\xi^p\}_{\xi \in \mathbf{Z}^n}, p = 0, 1, \ldots$, having the following properties (also see, e.g., [CoHi, ReSi, Mt]): for each nonnegative integer p, $\{Q_\xi^p\}_{\xi \in \mathbf{Z}^n}$ is a tessellation of $\mathbf{R}^n$ into a countable family of congruent and nonoverlapping open n-dimensional cubes of side $\epsilon_p = 2^{-p}$ such that $\cup_{\xi \in \mathbf{Z}^n} Q_\xi^p = \mathbf{R}^n$. Further, the cubes of the pth 'generation' are obtained by halving the sides of each cube of the previous generation. We define (by induction on p) the following index set I_p and open subsets Ω_p', ω_p of Ω:

$$I_0 = \{\xi \in \mathbf{Z}^n : Q_\xi^0 \subset \Omega\}, \quad \Omega_0' = \cup_{\xi \in I_0} Q_\xi^0, \qquad \omega_0 = \Omega \setminus \overline{\Omega_0'};$$
$$I_1 = \{\xi \in \mathbf{Z}^n : Q_\xi^1 \subset \omega_0\}, \quad \Omega_1' = \left(\cup_{\xi \in I_1} Q_\xi^1\right) \cup \Omega_0', \quad \omega_1 = \Omega \setminus \overline{\Omega_1'},$$

and so on. We observe that for all $p \geq 1$: $\omega_p \subset \Gamma_{c_1 \epsilon_p} \cap \Omega$, where $c_1 = 1 + \sqrt{n}$ and $\epsilon_p = 2^{-p}$.

Proposition 6.8 *There exist positive constants $\lambda_1 > 1, c_1$ and p_1 such that for all integers $p \geq p_1$ and all $\lambda \geq \lambda_1\epsilon_p^{-2}$, we have*

$$|N(\lambda) - \varphi(\lambda)| \leq c_1 \left(\epsilon_p^n \frac{1}{h(\epsilon_p)} \lambda^{\frac{n}{2}} + R_p(\lambda)\right), \tag{6.13}$$

where

$$R_p(\lambda) := \sum_{q=0}^{p} (\#I_q)(1 + \epsilon_q^{n-1}\lambda^{\frac{n-1}{2}}). \tag{6.14}$$

Proof. Fix $p > 0$ and $\lambda > 0$. Let $\Omega' = \Omega_p'$ and $\omega' = \omega_p$. Applying (6.4) with $W_0 = H_0^1(\Omega')$ and $W = H_0^1(\Omega)$, we have

$$N_0(\lambda; \Omega') \leq N(\lambda) \leq N_0(\lambda; \Omega') + \mathcal{N}(\lambda; \omega),$$

where $\mathcal{N}(\lambda; \omega) := N(\lambda; Z_\lambda, L^2(\Omega), a_\Omega)$; it follows that

$$A_1 + A_2 \leq N(\lambda) - \varphi(\lambda) \leq A_1 + A_2 + A_3, \tag{6.15}$$

where

$$
\begin{aligned}
A_1 &:= N_0(\lambda; \Omega') - \varphi(\lambda; \Omega'), \\
A_2 &:= \varphi(\lambda; \Omega') - \varphi(\lambda; \Omega), \\
A_3 &:= \mathcal{N}(\lambda; \omega).
\end{aligned}
$$

Since $\Omega' = \Omega'_p = \cup_{q=0}^{p}(\cup_{\xi \in I_q} Q_\xi^q)$, we have

$$
N_0(\lambda; \Omega') = \sum_{q=0}^{p} \sum_{\xi \in I_q} N_0(\lambda; Q_\xi^q), \quad \varphi(\lambda; \Omega') = \sum_{q=0}^{p} \sum_{\xi \in I_q} \varphi(\lambda; Q_\xi^q),
$$

and hence

$$
|A_1| \le \sum_{q=0}^{p} \sum_{\xi \in I_q} |N_0(\lambda; Q_\xi^q) - \varphi(\lambda; Q_\xi^q)|. \tag{6.16}
$$

We can thus apply (6.5) with $W_Q = H_0^1(Q)$ to estimate uniformly each summand in (6.16) and deduce that there exists a positive constant c such that for all $p \ge 1$ and $\lambda > 0$,

$$
|A_1| \le c \sum_{q=0}^{p} (\#I_q) \left(1 + \epsilon_q^{n-1} \lambda^{\frac{n-1}{2}}\right) = c R_p(\lambda). \tag{6.17}
$$

For A_2, we have

$$
\begin{aligned}
|A_2| &= \varphi(\lambda; \Omega) - \varphi(\lambda; \Omega') = (2\pi)^{-n} \mathcal{B}_n (|\Omega|_n - |\Omega'|_n) \lambda^{\frac{n}{2}} \\
&= (2\pi)^{-n} \mathcal{B}_n |\omega|_n \lambda^{\frac{n}{2}}.
\end{aligned} \tag{6.18}
$$

Next, we recall that $\omega = \omega_p \subset \Gamma_{c_1 \epsilon_p} \cap \Omega$, with $c_1 = 1 + \sqrt{n}$. Since $\epsilon_p = 2^{-p} \to 0$ and $M^*(h; \Gamma) = \limsup_{\epsilon \to 0^+} \epsilon^{-n} h(\epsilon) |\Gamma_\epsilon \cap \Omega|_n < \infty$, we deduce that there exist c_2 and p'_1 such that for all $p \ge p'_1$,

$$
|\omega|_n = |\omega_p|_n \le |\Gamma_{c_1 \epsilon_p} \cap \Omega|_n \le c_2 (c_1 \epsilon_p)^n \frac{1}{h(c_1 \epsilon_p)} \le c'_1 \epsilon_p^n \frac{1}{h(\epsilon_p)}.
$$

The last inequality follows from the fact that $h(c_1 \epsilon) \le h(\epsilon_p)$ since $c_1 > 1$ and h is nondecreasing. Thus (6.18) yields

$$
|A_2| \le c \epsilon_p^n \frac{1}{h(\epsilon_p)} \lambda^{\frac{n}{2}}, \quad \text{for all} \quad p \ge p'_1 \quad \text{and} \quad \lambda > 0. \tag{6.19}
$$

Finally, by [Mt, Lemma 5.8], there exists a positive constant c such that for all $\lambda > 0$, $\mathcal{N}(\lambda; \omega) \leq N_0^*(\lambda; \omega) + cR_p(\lambda)$. Next, we note that for all $\lambda > 0$, $N_0^*(\lambda; \omega) \leq N_1^*(\lambda; \omega) \leq N_1^*(\lambda; \Gamma_{c_1 \epsilon_p} \cap \Omega)$. We then apply (6.10) to deduce that there exist positive constants $\lambda_1 > 1, c_1'$ and $p_1 \geq p_1'$ such that for all $p \geq p_1$ and $\lambda \geq \lambda_1 \epsilon_p^{-2} (\geq 1)$, we have

$$A_3 = \mathcal{N}(\lambda; \omega) \leq c_1' \epsilon_p^n \frac{1}{h(\epsilon_p)} \lambda^{\frac{n}{2}} + cR_p(\lambda). \tag{6.20}$$

Now after putting (6.17), (6.19) and (6.20) into (6.15), we conclude that (6.13) holds. ∎

We can write

$$R_p(\lambda) = S_p \lambda^{\frac{n-1}{2}} + T_p, \tag{6.21}$$

where

$$S_p := \sum_{q=0}^{p} (\#I_q) \, \epsilon_q^{n-1} \quad \text{and} \quad T_p := \sum_{q=0}^{p} \#I_q. \tag{6.22}$$

Note that S_p and T_p are independent of λ.

Proposition 6.9 *There exist positive constants c_2, c_3, p_2 — depending only on Ω, n, d — such that the following inequality holds for all $p > p_2$:*

$$S_p \leq c_2 + c_3 \epsilon_p^{n-1} \frac{1}{h(\epsilon_p)}. \tag{6.23}$$

Proof. Recall that by construction

$$\cup_{\xi \in I_q} Q_\xi^q \subset \omega_{q-1} \subset \Gamma_{c_1 \epsilon_{q-1}} \cap \Omega, \text{ for all } q \geq 2,$$

with $c_1 = 1 + \sqrt{n}$. Using the assumption $M^*(h; \Gamma) < \infty$, we deduce that there exist constants c, c' and $p_2 > 1$, such that for all $q \geq p_2$,

$$
\begin{aligned}
| \cup_{\xi \in I_q} Q_\xi^q |_n &= (\#I_q) \epsilon_q^n \leq |\Gamma_{c_1 \epsilon_{q-1}} \cap \Omega|_n \\
&\leq c_1' (c_1 \epsilon_{q-1})^n \frac{1}{h(c_1 \epsilon_{q-1})} \leq c\epsilon_q^n \frac{1}{h(\epsilon_q)}.
\end{aligned}
$$

The last inequality follows from the fact that $\epsilon_{q-1} = 2\epsilon_q$ and $h(c_1 \epsilon_{q-1}) = h(2c_1 \epsilon_q) \geq h(\epsilon_q)$. So we have

$$\#I_q \leq c\frac{1}{h(\epsilon_q)}, \quad (\#I_q)\epsilon_q^{n-1} \leq c\epsilon_q^{n-1} \frac{1}{h(\epsilon_q)}, \tag{6.24}$$

for all $q \geq p_2$.

Next, we fix $p > p_2$ and break S_p into two sums :

$$S_p = \sum_{q=0}^{p}(\#I_q)\epsilon_q^{n-1} = \sum_{q=0}^{p_2}(\#I_q)\epsilon_q^{n-1} + \sum_{q=p_2+1}^{p}(\#I_q)\epsilon_q^{n-1}$$

$$\leq c_2 + c\sum_{q=p_2+1}^{p}\epsilon_q^{n-1}\frac{1}{h(\epsilon_q)}. \qquad (6.25)$$

By assumption (C2), we know that

$$k_1 h(x) \leq h(2x) \leq k_2 h(x)$$

with $2^{n-1} < k_1 \leq k_2 \leq 2^n$ for all $0 < x < x_0$, where x_0 is a small constant. Thus, without loss of generality, we can assume that p_2 is large enough so that $\epsilon_q = 2^{-q} < x_0$ for all $q > p_2$. Then for any $p > q > p_2$, we easily deduce that

$$\frac{1}{h(\epsilon_q)} = \frac{1}{h(2^{p-q}\epsilon_p)} \leq \frac{1}{k_1^{p-q}}\frac{1}{h(\epsilon_p)}.$$

Therefore, continuing with (6.25), we have for all $p > p_2$,

$$S_p \leq c_2 + c\frac{1}{h(\epsilon_p)}\sum_{q=p_2+1}^{p}\epsilon_q^{n-1}\frac{1}{k_1^{p-q}}$$

$$= c_2 + c\frac{\epsilon_p^{n-1}}{h(\epsilon_p)}\sum_{q=p_2+1}^{p}\frac{2^{(n-1)(p-q)}}{k_1^{p-q}}$$

$$\leq c_2 + c\left(\frac{1}{1-\frac{2^{n-1}}{k_1}}\right)\epsilon_p^{n-1}\frac{1}{h(\epsilon_p)}$$

$$= c_2 + c_3\epsilon_p^{n-1}\frac{1}{h(\epsilon_p)},$$

which completes the proof of this proposition. Note that in the last inequality, we have used the fact that for $\kappa \ (= k_1/2^{n-1}) > 1$,

$$\sum_{q=p_2+1}^{p}\kappa^{q-p} \leq \sum_{q=0}^{p}\kappa^{q-p} = \kappa^{-p}\frac{\kappa^{p+1}-1}{\kappa-1}$$

$$= \frac{\kappa - \kappa^{-p}}{\kappa - 1} \leq \frac{\kappa}{\kappa - 1} = \frac{1}{1-\kappa^{-1}}. \qquad \blacksquare$$

Proposition 6.10 *There exist positive constants c_2, c_3 and p_2 such that for all $p > p_2$,*

$$T_p \le c_2 + c_3 \frac{1}{h(\epsilon_p)}. \tag{6.26}$$

Proof. Using (6.24), we have since $k_1 > 1$,

$$
\begin{aligned}
T_p &= \sum_{q=0}^{p_2} \#I_q + \sum_{q=p_2+1}^{p} \#I_q \\
&\le c_2 + c \sum_{q=p_2+1}^{p} \frac{1}{h(\epsilon_q)} \\
&\le c_2 + c \frac{1}{h(\epsilon_p)} \sum_{q=p_2+1}^{p} \frac{1}{k_1^{p-q}} \le c_2 + c \frac{1}{h(\epsilon_p)} \frac{1}{1 - \frac{1}{k_1}} \\
&= c_2 + c_3 \frac{1}{h(\epsilon_p)},
\end{aligned}
$$

as desired. ∎

We can now conclude our proof of Theorem 2.12.

Putting (6.23) and (6.26) into (6.21), we see that there exist positive constants c_2, c_3 and p_2 such that for all $p > p_2$,

$$R_p(\lambda) \le (c_2 + c_3 \epsilon_p^{n-1} \frac{1}{h(\epsilon_p)}) \lambda^{\frac{n-1}{2}} + c_2 + c_3 \frac{1}{h(\epsilon_p)}. \tag{6.27}$$

Thus assuming that $p_2 > p_1$, and combining (6.27) and (6.13), we obtain that for all $p > p_2$ and $\lambda \ge \lambda_1 \epsilon_p^{-2}$:

$$
\begin{aligned}
|N(\lambda) - \varphi(\lambda)| &\le c_1' \left(\epsilon_p^n \frac{1}{h(\epsilon_p)} \lambda^{\frac{n}{2}} + \lambda^{\frac{n-1}{2}} + \epsilon_p^{n-1} \frac{1}{h(\epsilon_p)} \lambda^{\frac{n-1}{2}} + 1 + \frac{1}{h(\epsilon_p)} \right) \\
&=: c_1'(A + B + C + D),
\end{aligned} \tag{6.28}
$$

where

$$
\begin{aligned}
A &:= \epsilon_p^n \frac{1}{h(\epsilon_p)} \lambda^{\frac{n}{2}}, & B &:= \lambda^{\frac{n-1}{2}}, \\
C &:= \epsilon_p^{n-1} \frac{1}{h(\epsilon_p)} \lambda^{\frac{n-1}{2}}, & D &:= 1 + \frac{1}{h(\epsilon_p)}.
\end{aligned}
$$

Next we want to estimate each of these terms as $\epsilon \to 0^+$. We claim that if we choose p such that $2^{-p} = \epsilon_p \asymp \lambda^{-1/2}$, that is, $2^p = \epsilon_p^{-1} \asymp \lambda^{1/2}$ as $\lambda \to \infty$, then for all λ large enough, each term will have an upper bound of the form $cf(\sqrt{\lambda})$, as desired.

More explicitly, let $p = p(\lambda) := [\frac{1}{2} \log_2 (\lambda/\lambda_1)]$. Since $2^p \leq (\lambda/\lambda_1)^{1/2}$ and $2^{p+1} > (\lambda/\lambda_1)^{1/2}$, there exist positive constants a_1, a_2 such that

$$a_2^{-1}\lambda^{1/2} \leq 2^p = \epsilon_p^{-1} < a_1^{-1}\lambda^{1/2};$$

that is,

$$a_1\lambda^{-1/2} < \epsilon_p \leq a_2\lambda^{-1/2}. \tag{6.29}$$

Now let λ_2 be so large that $\lambda_2 > \lambda_1$ and $p(\lambda_2) > p_2$.

From the relation $k_1 h(x) \leq h(2x) \leq k_2 h(x)$, we can easily deduce that for any $M > 0$:

$$K_1(M)h(x) \leq h(Mx) \leq K_2(M)h(x),$$

for all x small enough, where $K_1(M), K_2(M)$ denote positive constants depending only on M. Thus we see that by (6.29) and since h is nondecreasing, there exist some positive constants b_1, b_2 such that for all λ large enough:

$$b_1 f(\sqrt{\lambda}) = \frac{b_1}{h(\lambda^{-1/2})} \leq \frac{1}{h(\epsilon_p)} \leq \frac{b_2}{h(\lambda^{-1/2})} = b_2 f(\sqrt{\lambda});$$

that is,

$$\frac{1}{h(\epsilon_p)} \asymp f(\sqrt{\lambda}), \quad \text{as} \ \lambda \to \infty.$$

Thus for all $\lambda > \lambda_2$:

$$A = \epsilon_p^n \frac{1}{h(\epsilon_p)}\lambda^{n/2} \asymp \lambda^{-n/2}\frac{1}{h(\epsilon_p)}\lambda^{n/2} \ (\leq b_2 f(\sqrt{\lambda})).$$

Similarly, we obtain

$$C = \epsilon_p^{n-1}\frac{1}{h(\epsilon_p)}\lambda^{\frac{n-1}{2}} \leq cf(\sqrt{\lambda}).$$

Further,

$$B = \lambda^{\frac{n-1}{2}} \leq cf(\sqrt{\lambda}), \quad \text{for all } \lambda \ \text{large enough.}$$

(This follows from hypothesis (C3) which says that $h(x) \leq cx^{n-1}$ for all x small, and thus $cf(x) \geq x^{n-1}$ for all x large enough.)

As for the last term, we have:

$$D = 1 + \frac{1}{h(\epsilon_p)} \leq 2\frac{1}{h(\epsilon_p)} \leq cf(\sqrt{\lambda}).$$

Now if we add all these estimates together, we have, in view of (6.28),

$$|N(\lambda) - \varphi(\lambda)| \leq Cf(\sqrt{\lambda}), \quad \text{for all } \lambda \text{ large enough};$$

that is,

$$N(\lambda) = \varphi(\lambda) + O(f(\sqrt{\lambda})), \quad \text{as } \lambda \to \infty,$$

as desired.

This completes our proof of Theorem 2.12. ∎

Remark 6.11 (a) Theorem 2.12 is also valid when $n = 1$ (under the same assumptions (C1)—(C4) on h). Indeed, the above proof holds without change in that situation.

(b) As mentioned previously, in view of the results in [La1], Theorem 2.12 also extends (under suitable hypotheses) without difficulty to higher order operators and to Neumann (or more generally, mixed) boundary conditions. Given our work in the present section, the statements and proofs parallel those of [La1, Theorem 2.1, p. 479] and [La1, Theorem 4.1, pp. 510-511], respectively. For example, for the Neumann Laplacian, we assume that Ω satisfies the so-called 'extension property' (which is the case if $\Omega \subset \mathbf{R}^2$ is a bounded simply connected quasidisk [Jo; Ma, p. 70]) and we replace one-sided by two-sided neighborhoods in Definition 2.1 of $M^*(h; \Gamma)$ [i.e., if we substitute Γ_ϵ for $\Gamma_\epsilon \cap \Omega$ in (2.1)]; see [La1, pp. 510-511] for more details. (Recall that $\Omega \subset \mathbf{R}^n$ is said to satisfy the 'extension property' if there exists a continuous linear map $\mathcal{E} : H^1(\Omega) \to H^1(\mathbf{R}^n)$ extending every function in the Sobolev space $H^1(\Omega)$ to a function in $H^1(\mathbf{R}^n)$.)

7 Examples

In this section, we illustrate our results by various examples, both in the one and higher dimensional cases.

We first consider, in Examples 7.1 and 7.3, the case when $n = 1$. We begin with a very simple example illustrating Corollary 2.6.

Example 7.1 For any $h \in G_d$ with $d \in (0,1)$, let $l_j = g(j) = h^{-1}(1/j)$ for all $j \geq 1$ and let Ω be an open subset of $\mathbf{R}$ (necessarily of finite length $|\Omega|_1 = \sum_{j=1}^{\infty} l_j < \infty$) having $(l_j)_{j=1}^{\infty}$ for associated sequence. Then, by Theorems 2.4 and 2.5, $\Gamma = \partial\Omega$ is h-Minkowski measurable with h-Minkowski content $M(h; \Gamma) = \frac{2^{1-d}}{1-d}$ (where $L = 1$), and the asymptotic expansion of $N(\lambda)$ admits a *monotonic* second term; more precisely, we have

$$N(\lambda) = \varphi(\lambda) + \pi^{-d}\zeta(d)f(\sqrt{\lambda}) + o(f(\sqrt{\lambda})), \quad \text{as} \quad \lambda \to \infty,$$

where $\varphi(\lambda) = \pi^{-1}|\Omega|_1\lambda^{1/2}$ and $f(x) = 1/h(1/x)$.

Remark 7.2 According to Example 2.3, we may take, for instance, $h(x) = x^d(\ln(x^{-1}+1))^{-a}$, with $a \geq 0$, and so $f(x) := x^d(\ln(x+1))^a$. (In view of Remark 2.10, this comment will also apply to Example 7.3 below.)

The next example will show, in particular, that the assumption of h-Minkowski measurability of Γ in Corollary 2.6 is indeed necessary.

Example 7.3 We still assume that $n = 1$ and $d \in (0,1)$. Recall that in Section 5, given $h \in G_d$, with h differentiable and $xh'(x)/h(x) \geq \mu > 0$, where μ is a constant, we constructed a positive decreasing sequence $(l_j)_{j=1}^{\infty}$ (see (5.1)—(5.3)) such that $l_j/g(j)$ does not converge, but as in (5.21),

$$
\begin{aligned}
\delta(x) \;&:=\; \sum_{j=1}^{\infty}\{l_j x\} = \sum_{j=1}^{\infty} l_j x - \sum_{j=1}^{\infty}[l_j x] = \left(\sum_{j=1}^{\infty} l_j\right) x - \mathcal{N}(x) \\
&=\; -\zeta(d)f(x) - \beta\zeta(d+i\nu)f(x)x^{i\nu} - \beta\zeta(d-i\nu)f(x)x^{-i\nu} + o(f(x)).
\end{aligned}
$$

In other words,

$$\mathcal{N}(x) = (\sum_{j=1}^{\infty} l_j)x + \zeta(d)f(x) + \beta\zeta(d+i\nu)f(x)x^{i\nu}$$

$$+ \beta\zeta(d-i\nu)f(x)x^{-i\nu} + o(f(x)), \text{ as } x \to \infty. \qquad (7.1)$$

Now let $\Omega \subset \mathbf{R}$ be an open set (necessarily of finite length) having $(l_j)_{j=1}^{\infty}$ for associated sequence. Hence, by Theorem 2.5, $\Gamma = \partial\Omega$ is *not* Minkowski measurable. Further, by letting $x = \sqrt{\lambda}/\pi$ in (7.1), we see that as $\lambda \to \infty$:

$$\begin{aligned} N(\lambda) &= \varphi(\lambda) + \pi^{-d}\zeta(d)f(\sqrt{\lambda}) + \beta\zeta(d+i\nu)\pi^{-(d+i\nu)}\lambda^{\frac{i\nu}{2}}f(\sqrt{\lambda}) \\ &\quad + \beta\zeta(d-i\nu)\pi^{-(d-i\nu)}\lambda^{\frac{-i\nu}{2}}f(\sqrt{\lambda}) + o(f(\sqrt{\lambda})) \\ &= \varphi(\lambda) + \pi^{-d}f(\sqrt{\lambda})\left(\zeta(d) + 2\beta Re\left(\zeta(d+i\nu)\pi^{-i\nu}\lambda^{\frac{i\nu}{2}}\right)\right) + o(f(\sqrt{\lambda})). \end{aligned}$$
$$(7.2)$$

(This follows since by Lemma 3.1(1), $f(\sqrt{\lambda}/\pi) = \pi^{-d}f(\sqrt{\lambda})(1 + o(1))$ as $\lambda \to \infty$.)

In the proof of Theorem 5.1, we assumed that $d + i\nu$ (and hence also $d - i\nu$) is a zero of the Riemann zeta-function; thus in that case, we obtain

$$N(\lambda) = \varphi(\lambda) + \pi^{-d}\zeta(d)f(\sqrt{\lambda}) + o(f(\sqrt{\lambda})), \text{ as } \lambda \to \infty.$$

But now if we choose $\nu \in \mathbf{R}$ in such a way that $d + i\nu$ is *not* a zero of $\zeta(s)$ (see Remark 7.4(a) below), then we deduce from (7.2) that (for $\beta > 0$ small enough) the counting function $N(\lambda)$ does not have a monotonic second term, but the coefficient of $f(\sqrt{\lambda})$ is a (nontrivial) periodic function of $\ln\lambda$, instead of a constant. (See Remark 5.6(b).) Therefore, the assumption we have made about the h-Minkowski measurability of the boundary $\Gamma = \partial\Omega$ in Corollary 2.6 is necessary. Note that this example does not contradict Theorem 2.7 because, as is easily checked, $0 < M_* < M^* < \infty$ (since by construction $l_j \asymp g(j)$) and $\varphi(\lambda) - N(\lambda) \asymp f(\sqrt{\lambda})$.

Remark 7.4 (a) Given any $d \in (0,1)$, we can always find $\nu \in \mathbf{R}$ as above because $\zeta = \zeta(s)$ being analytic in the 'critical strip' $0 < Re\ s < 1$, it has at most countably many zeros on the vertical line $Re\ s = d$. Of course, this is the case whether or not the Riemann hypothesis is true.

(b) In agreement with the intuitive comments made in [LaPo1-2, LaMa, La3] (see Remark 5.6(c) above), the *oscillations* in the asymptotic second term of $N(\lambda)$ are caused by $\pm\nu$, the imaginary part of the complex exponent $d\pm i\nu$.

(c) In [LaPo2, Example 4.5, pp. 65-67] (or [LaPo1]), another example (really of a different nature) is given of an open set $\Omega \subset \mathbf{R}$ such that $\Gamma = \partial\Omega$ is not h-Minkowski measurable (with $h(x) = x^d$) but $N(\lambda)$ admits an oscillatory asymptotic second term (while at the same time, $0 < M_*(h;\Gamma) \leq M^*(h;\Gamma) < \infty$) and $\varphi(\lambda) - N(\lambda) \asymp \lambda^{d/2}$, as $\lambda \to +\infty$). Namely, let $\Omega = (0,1) \setminus K$, where K is the classical ternary Cantor set (of Minkowski dimension $d = \ln 2/\ln 3$); then, clearly, the boundary Γ of Ω is equal to K. It would be interesting to find variants of this Cantor set example (perhaps by using some of the perfect, symmetric compact sets studied in [KaSa]) corresponding to diverse types of gauge functions.

Next we turn our attention to the higher dimensional case. We are going to use the following examples to show that our estimate in Theorem 2.12 really improves upon the previous results of the second author in [La1] and is in general, best possible.

Example 7.5 In [EvHa, §6.2, pp. 515-526], Evans and Harris constructed a (simply connected) domain Ω in $\mathbf{R}^2$ from a succession of finite sets (generations) Δ_m of closed congruent rectangles Q_m of edge length $\alpha_m = 1/4^m$, $\beta_m = 1/2^m$ and with disjoint interiors; see Figure 1. The generation Δ_0 consists of a single rectangle as does Δ_1, a short edge of Δ_1 being attached to the middle portion of a long edge of Q_0. For $m \geq 1$, Δ_m contains 2^{m-1} rectangles and to each long edge of Q_m is attached a short edge of a rectangle Q_{m+1} belonging to the next generation. The domain Ω is the interior of the connected set Δ constructed in this way; that is,

$$\Omega = \Delta^0,$$

where

$$\Delta = \bigcup_{m\in\mathbf{N}} \{Q_m : Q_m \in \Delta_m\}.$$

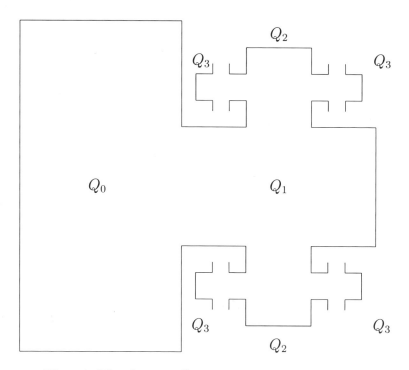

Fig. 1 *The domain Ω: rooms and passages*

The following estimate was obtained in [EvHa, p. 519]:

$$|\Gamma_\epsilon \cap \Omega|_2 \asymp \epsilon \ln 1/\epsilon, \quad \text{as} \quad \epsilon \to 0^+. \tag{7.3}$$

Now let

$$h(x) = \frac{x}{\ln(\frac{1}{x} + 1)}. \tag{7.4}$$

It can be easily checked that this function h satisfies all the assumptions (C1)—(C4) in Section 2.2. Further, (7.3) implies that

$$\epsilon^{-2} h(\epsilon) |\Gamma_\epsilon \cap \Omega|_2 \asymp \epsilon^{-2} \frac{\epsilon}{\ln(1/\epsilon + 1)} \epsilon \ln 1/\epsilon \asymp 1, \quad \text{as} \quad \epsilon \to 0^+;$$

so taking both the upper and lower limit, we obtain in view of Definition 2.1 of the upper and lower h-Minkowski contents,

$$0 < M_*(h; \Gamma) \leq M^*(h; \Gamma) < \infty.$$

Thus we deduce from Theorem 2.12 that as $\lambda \to \infty$:

$$N(\lambda) - \varphi(\lambda) = O(f(\sqrt{\lambda}) = O(\lambda^{1/2} \ln(\sqrt{\lambda} + 1)) = O(\lambda^{1/2} \ln \lambda), \qquad (7.5)$$

which is exactly what was found in [EvHa] by means of a direct computation.

Remark 7.6 In view of (7.3) and Definition 1.1, the boundary $\Gamma = \partial\Omega$ has Minkowski dimension $D = 1$ and D-dimensional upper Minkowski content $M^*(D; \Gamma) = \infty$. (Recall that in Definition 1.1, we use the standard 'gauge function' $h(x) = x^D$, which is just x in this case.) Hence, in the terminology of [La1], we are in the 'nonfractal case' where $D = n - 1 = 1$. However (as noted in [EvHa]), since $M^*(D; \Gamma) = \infty$, [La1, Theorem 2.1, p. 479] only yields as $\lambda \to \infty$,

$$N(\lambda) - \varphi(\lambda) = O(\lambda^{1/2+\epsilon}), \quad \text{for all} \quad \epsilon > 0.$$

Therefore, the use of the 'gauge function' h in (7.4) combined with Theorem 2.12 enables us to improve upon the results of [La1] in this situation.

It would be interesting to show that, as we conjecture, the estimate (7.5) above is actually sharp.

The following example is contained in [Ce1]. It will show that our estimate in Theorem 2.12 is optimal.

Example 7.7 In [Ce1], Caetano constructed the following open set in $\mathbf{R}^2$, which extends Examples 5.1 and 5.1' in [La1] that were used to show that the estimates of [La1] (recalled in Theorem 1.2) are sharp.

Let $a(x) = x^{-1} \ln x$. Note that a is strictly decreasing on $(3, \infty)$. Define the bounded open subset Ω of $\mathbf{R}^2$ by

$$\Omega := \cup_{j=3}^{\infty}(a(j + 1), a(j)) \times (-\frac{1}{2}, \frac{1}{2}).$$

See Figure 2.

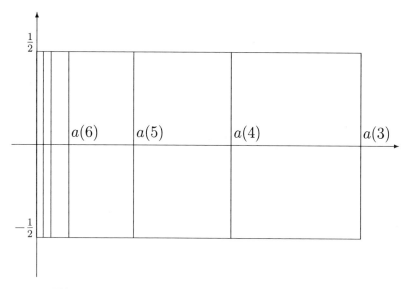

Fig. 2 *The open set Ω: fractal comb*

Let $b(x) = \ln x$, which is positive on $(3, \infty)$ and satisfies $\lim_{x \to \infty} x b'(x)/b(x) = 0$. Further, let $r(x) = x^{-2} \ln x$, and denote by $r^{-1}(x)$ its inverse function.

By Proposition 2.10 in [Ce1], there exist positive constants c_1 and c_2 such that

$$c_1 \epsilon r^{-1}(\epsilon) \le |\Gamma_\epsilon \cap \Omega|_2 \le c_2 \epsilon r^{-1}(\epsilon), \quad \text{for all small } \epsilon. \qquad (7.6)$$

Using Proposition 2.11 in [Ce1], we have

$$r^{-1}(\lambda^{-1/2}) \asymp \lambda^{1/4}(b(\lambda))^{1/2}, \quad \text{as} \quad \lambda \to \infty.$$

If we let $\epsilon = \lambda^{-1/2}$, we obtain that

$$r^{-1}(\epsilon) \asymp \epsilon^{-1/2}(b(\epsilon^{-2}))^{1/2}, \quad \text{as} \quad \epsilon \to 0^+;$$

that is,

$$\begin{aligned} r^{-1}(\epsilon) &\asymp \epsilon^{-1/2}(\ln \epsilon^{-2})^{1/2} \\ &\asymp \epsilon^{-1/2}(\ln \epsilon^{-1})^{1/2}, \quad \text{as} \quad \epsilon \to 0^+. \end{aligned} \qquad (7.7)$$

Putting (7.7) into (7.6), we see that

$$c_1 \epsilon^{1/2}(\ln \epsilon^{-1})^{1/2} \le |\Gamma_\epsilon \cap \Omega|_2 \le c_2 \epsilon^{1/2}(\ln \epsilon^{-1})^{1/2}. \qquad (7.8)$$

Now, let

$$h(x) = \frac{x^{3/2}}{(\ln(x^{-1} + 1))^{1/2}}. \tag{7.9}$$

We can easily check much like in Example 3 of the appendix that the function h satisfies all the assumptions (C1)—(C4).

It follows from (7.8) that

$$\epsilon^{-2}h(\epsilon)|\Gamma_\epsilon \cap \Omega|_2 \le c_2\epsilon^{1/2}(\ln \epsilon^{-1})^{1/2}\epsilon^{-2}\frac{\epsilon^{3/2}}{(\ln(\epsilon^{-1} + 1))^{1/2}}.$$

So if we take the upper limit as $\epsilon \to 0^+$, we deduce from (2.1) that

$$M^*(h; \Gamma) \le c_2 < \infty.$$

(Of course, we also have $M_*(h; \Gamma) \ge c_1 > 0$.) Therefore, by our Theorem 2.12, it follows that

$$\begin{aligned} N(\lambda) - \varphi(\lambda) &= O(f(\sqrt{\lambda})) = O(\lambda^{3/4}(\ln(\sqrt{\lambda} + 1))^{1/2}) \\ &= O(\lambda^{3/4}(\ln \lambda)^{1/2}), \quad \text{as} \quad \lambda \to \infty. \end{aligned} \tag{7.10}$$

We actually obtain a sharp estimate in this case; indeed, by Corollary 2.15 in [Ce1], we have

$$N(\lambda) - \varphi(\lambda) \asymp f(\sqrt{\lambda}) \asymp \lambda^{3/4}(\ln \lambda)^{1/2}, \quad \text{as} \quad \lambda \to \infty. \tag{7.11}$$

(Note that by (7.9),

$$f(x) = \frac{1}{h(1/x)} = x^{3/2}(\ln(x + 1))^{1/2}$$

and so, as was used in (7.10) and (7.11),

$$f(\sqrt{\lambda}) = \lambda^{3/4}(\ln(\sqrt{\lambda} + 1)^{1/2} \asymp \lambda^{3/4}(\ln \lambda)^{1/2},$$

as $\lambda \to \infty$.)

Remark 7.8 (a) It follows easily from (7.8) and Definition 1.1 that

$$M_*(d; \Gamma) = M^*(d; \Gamma) = \begin{cases} \infty & \text{if } d \le 3/2, \\ 0 & \text{if } d > 3/2. \end{cases}$$

Hence, $\Gamma = \partial\Omega$ has *Minkowski dimension* $D = D(\Gamma) = 3/2$ and $M_*(D;\Gamma) = M^*(D;\Gamma) = \infty$. Clearly— as noted in [La1, Remark 2.4(e), p. 481]—in such a situation, $h(x) = x^D$ is *not* the proper 'gauge function'; in fact, for the present example, $h(x) = \frac{x^D}{(\ln(1+x^{-1}))^{1/2}}$ (as given by (7.9)) is an '*admissible gauge function*', both from the geometric and spectral points of view.

(b) Analogous results hold for the entire family of planar domains studied in [Ce1] (involving various functions a, b and r). Further, as in [La1, Examples 5.1 and 5.1', pp. 511-515], such examples providing sharp error estimates can clearly be constructed in any dimension $n \geq 1$ and with $D = D(\Gamma)$ ranging in all of the interval $(n-1, n)$.

(c) In our various examples, we could also include gauge functions of the form a power function times a logarithm.

For instance, in Example 7.7, we can let $a(x) = 1/(x \ln x)$, $b(x) = 1/\ln x$ and $r(x) = 1/(x^2 \ln x)$. Define the gauge function

$$h(x) = x^{3/2}(\ln(1/x + 1))^{1/2}.$$

It can be easily checked that this h also satisfies (C1)—(C4). Now proceeding as in (7.6), (7.7) and (7.8), we obtain that

$$0 < M_*(h;\Gamma) \leq M^*(h;\Gamma) < \infty.$$

Thus it follows from Theorem 2.12 that as $\lambda \to \infty$,

$$N(\lambda) - \varphi(\lambda) = O(f(\sqrt{\lambda})) = O(\frac{\lambda^{3/4}}{(\ln \lambda)^{1/2}}).$$

We also obtain a sharp estimate, since by Corollary 2.15 in [Ce1], we have

$$\varphi(\lambda) - N(\lambda) \asymp \frac{\lambda^{3/4}}{(\ln \lambda)^{1/2}}.$$

Further, in this case, Γ has Minkowski dimension $D = 3/2$ and $M_*(D;\Gamma) = M^*(D;\Gamma) = 0$.

Appendix: Examples of Gauge Functions

We are going to carry out some detailed calculations to show that there is a large class of functions satisfying hypothesis (H1)—(H3) when $n = 1$ (see Examples 1 and 2 below), as well as hypothesis (C1)—(C4) when $n \geq 1$ (see Examples 3 and 4).

Example 1 Assume that $n = 1$. Set, as in Example 2.3,

$$h(x) = \frac{x^d}{\left(\ln(\frac{1}{x} + 1)\right)^a},$$

for some $0 < d < 1$ and $a \geq 0$. We claim that $h \in G_d$.

(H1). It is easy to see that h is positive, strictly increasing for $x > 0$, and $\lim_{x \to 0+} h(x) = 0$, $\lim_{x \to \infty} h(x) = \infty$, $\lim_{x \to 0+} h(x)/x = \infty$.

(H2). For any $t > 0$, $t \in [u, v] \subset (0, \infty)$,

$$\lim_{x \to 0+} \frac{h(tx)}{h(x)} = \lim_{x \to 0+} \frac{t^d x^d}{\left(\ln(\frac{1}{tx} + 1)\right)^a} \frac{\left(\ln(\frac{1}{x} + 1)\right)^a}{x^d}$$

$$= \lim_{x \to 0+} t^d \left(\frac{\ln \frac{1}{x}}{\ln \frac{1}{t} + \ln \frac{1}{x}} \right) = t^d.$$

Since t^d, $\ln t$ are all bounded on the compact set $[u, v]$, it follows that the convergence is uniform on any compact subset of $(0, \infty)$.

(H3). We want to show that

$$\frac{h(tx)}{h(x)} = \frac{t^d x^d}{\left(\ln(\frac{1}{tx} + 1)\right)^a} \frac{\left(\ln(\frac{1}{x} + 1)\right)^a}{x^d} \geq m t^\tau$$

for all x, t small, where $0 < \tau < 1$ and m is a positive constant; i.e., for all x, t small,

$$\frac{\ln(\frac{1}{x} + 1)}{\ln(\frac{1}{tx} + 1)} \geq \left(m t^{\tau - d} \right)^{1/a} = m^{1/a} t^{\frac{\tau - d}{a}},$$

or equivalently,

$$\frac{\ln \frac{1}{x}}{\ln \frac{1}{tx}} = \frac{1}{\frac{\ln(1/t)}{\ln(1/x)} + 1} \geq c t^{\frac{\tau - d}{a}}.$$

Since

$$\frac{1}{\frac{\ln(1/t)}{\ln(1/x)} + 1} > \frac{1}{\ln(1/t) + 1}, \quad \text{for all} \quad x < 1/e,$$

it suffices to show that

$$\frac{1}{\ln(1/t) + 1} > ct^b$$

for some $b > 0$ and all small t.

If we let $s = 1/t$, this is equivalent to showing that $\ln s + 1 < cs^b$ for all large s. This certainly can be done; for example, we can take $c = 2$. Indeed, let $w(s) = 2s^b - \ln s - 1$ for $s > 1$. Then $w(1) = 1 > 0$ and $w'(s) = 2bs^{b-1} - 1/s = 2bs^{-1}(s^b - 1) > 0$. Thus $w(s)$ is an increasing function for $s > 1$ and hence $w(s) > w(1) > 0$ for all $s > 1$. That is, $2s^b > \ln s + 1$. Now let $\frac{\tau - d}{a} = b$; then $\tau = ab + d$. So we can choose b small enough (e.g., $b < \frac{1-d}{a}$) to make $0 < \tau < 1$. Thus we showed that $\frac{h(tx)}{h(x)} \geq mt^\tau$ for all $0 < t < 1$ and $x < 1/e$, for some constants $m > 0, 0 < \tau < 1$.

Consequently, we have verified that this function h belongs to G_d. Also, we can show that it satisfies the additional assumption made in Theorem 2.9, as claimed in Remark 2.10. Note that

$$\lim_{x \to 0^+} \frac{xh'(x)}{h(x)} = \lim_{x \to 0^+} \frac{dx^d \left(\ln(\frac{1}{x} + 1) \right)^a + ax^d \left(\ln(\frac{1}{x} + 1) \right)^{a-1} \frac{1}{x+1}}{x^d \left(\ln(\frac{1}{x} + 1) \right)^a}$$

$$= \lim_{x \to 0^+} \left(d + a \left(\ln(\frac{1}{x} + 1) \right)^{-1} \frac{1}{x+1} \right) = d.$$

So there exists $x_1 > 0$ such that for all $0 < x < x_1$, we have $\frac{xh'(x)}{h(x)} \geq \frac{d}{2}$. Similarly, since

$$\lim_{x \to \infty} \frac{xh'(x)}{h(x)} = \lim_{x \to \infty} \left(d + a \left(\ln(\frac{1}{x} + 1) \right)^{-1} \frac{1}{x+1} \right) = d + a,$$

there exists x_2 such that for all $x > x_2$, we have $\frac{xh'(x)}{h(x)} \geq \frac{d}{2}$. Now on the compact interval $[x_1, x_2]$, $\frac{xh'(x)}{h(x)} > 0$; thus there exists $\eta > 0$ such that $\frac{xh'(x)}{h(x)} \geq \eta > 0$ on $[x_1, x_2]$.

Let $\mu = \min(\eta, d/2)$; then we see that

$$\frac{xh'(x)}{h(x)} \geq \mu > 0, \quad \text{for all } x > 0,$$

as desired. ∎

The next example will be a little bit more complicated.

Example 2 We still assume that $n = 1$. Set, as in Example 2.3,

$$h(x) = \frac{x^d}{\left(\ln(\ln(\frac{1}{x} + 1) + 1)\right)^a},$$

for some $0 < d < 1$ and $a \geq 0$. We want to verify that $h \in G_d$.

(H1). Clearly, $h : (0, \infty) \to (0, \infty)$ is positive, strictly increasing and $\lim_{x \to 0^+} h(x) = 0$, $\lim_{x \to \infty} h(x) = \infty$, $\lim_{x \to 0^+} h(x)/x = \infty$.

(H2). We have

$$\lim_{x \to 0^+} \frac{h(tx)}{h(x)} = \lim_{x \to 0^+} \left(\frac{\ln(\ln \frac{1}{x})}{\ln(\ln \frac{1}{tx})} \right)^a = t^d,$$

and since t^d, $\ln 1/t$ are bounded on any compact subset of $(0, \infty)$, we see that the convergence is uniform on such subsets.

(H3).
$$\frac{h(tx)}{h(x)} = t^d \left(\frac{\ln(\ln(\frac{1}{x} + 1) + 1)}{\ln(\ln(\frac{1}{tx} + 1) + 1)} \right)^a.$$

So
$$\frac{h(tx)}{h(x)} \geq mt^\tau, \quad \text{for all } x, t \text{ small}$$

if and only if

$$\frac{\ln(\ln(\frac{1}{x} + 1) + 1)}{\ln(\ln(\frac{1}{tx} + 1) + 1)} \geq m^{\frac{1}{a}} t^{\frac{\tau - d}{a}}.$$

Thus if we can prove that

$$\frac{\ln \ln \frac{1}{x}}{\ln \ln \frac{1}{tx}} \geq ct^b,$$

for some $b > 0$ and all x, t small, we are done. Now

$$\frac{\ln(\ln \frac{1}{x})}{\ln(\ln \frac{1}{tx})} = \frac{\ln(\ln \frac{1}{x})}{\ln(\ln \frac{1}{t} + \ln \frac{1}{x})} = \frac{\ln(\ln \frac{1}{x})}{\ln \ln \frac{1}{x} + \ln \left(1 + \frac{\ln \frac{1}{t}}{\ln \frac{1}{x}}\right)}$$

$$= \frac{1}{1 + \frac{\ln(1 + \ln 1/t / \ln 1/x)}{\ln \ln 1/x}} > \frac{1}{1 + \ln \left(1 + \frac{\ln(1/t)}{e}\right)},$$

if $x < e^{-e}$. Since

$$\frac{1}{1 + \ln\left(1 + \frac{\ln(1/t)}{e}\right)} \sim \frac{1}{\ln\ln\frac{1}{t}}, \quad \text{as } t \to 0^+,$$

it will be sufficient to show that $\frac{1}{\ln\ln(1/t)} > ct^b$ for some $b > 0$ and all t small. This is equivalent to showing that

$$\ln\ln\frac{1}{t} < ct^{-b}, \quad \text{for all } t \text{ small},$$

which is the same as saying that

$$\ln\ln s < cs^b, \quad \text{for all } s = \frac{1}{t} \text{ large}.$$

This can be shown. For example, consider $s > e, c = 1$. Then $s^b > \ln s > \ln\ln s$ for all $s > e$ and some $b > 0$. So $\ln\ln\frac{1}{t} < ct^{-b}$ for $t < e^{-1}$. Therefore,

$$\frac{\ln\ln\frac{1}{x}}{\ln\ln\frac{1}{tx}} > ct^b, \quad \text{for all } t \text{ small}.$$

Thus

$$\frac{h(tx)}{h(x)} \geq c't^\tau, \quad \text{for all } t \text{ small},$$

where $\tau = d + ab$. (So we should choose $b > 0$ small enough to make $\tau < 1$.)

We can now conclude that $h \in G_d$. It is not hard to verify that h also satisfies the additional assumption made in Theorem 2.9. After some simplifications, we obtain that

$$\frac{xh'(x)}{h(x)} = d + \frac{a}{\ln(\ln(\frac{1}{x} + 1) + 1)(\ln(\frac{1}{x} + 1) + 1)(x + 1)}.$$

So, $\lim_{x \to 0+} \frac{xh'(x)}{h(x)} = d$ and $\lim_{x \to \infty} \frac{xh'(x)}{h(x)} = d + a > d$. Hence there exist x_1, x_2 such that for all $0 < x < x_1$, we have $\frac{xh'(x)}{h(x)} \geq \frac{d}{2}$, and for all $x > x_2$, we have $\frac{xh'(x)}{h(x)} \geq \frac{d}{2}$. On the compact set $[x_1, x_2]$, $\frac{xh'(x)}{h(x)} > 0$. Thus there exists $\eta > 0$ such that $\frac{xh'(x)}{h(x)} \geq \eta > 0$. Let $\mu = \min(\frac{d}{2}, \eta)$; then for all $x > 0$,

$$\frac{xh'(x)}{h(x)} \geq \mu.$$

Consequently, we see that there are many functions which satisfy all the assumptions that we have made in Section 2.1.

Example 3 We now assume that $n \geq 1$. Set, as in Example 3.10,

$$h(x) = \frac{x^d}{(\ln(\frac{1}{x} + 1))^a},$$

for some $d \in (n - 1, n)$ and $a \geq 0$.

We can easily check that this function h satisfies all the assumptions (C1)—(C4) made in Section 2.2:

(C1). It is obvious that $h : (0, \infty) \to (0, \infty)$ is a strictly increasing function and $\lim_{x \to 0+} h(x) = 0$, $\lim_{x \to \infty} h(x) = \infty$.

(C2). Since $\lim_{x \to 0+} h(2x)/h(x) = 2^d$, there exist constants c_1, c_2, with $2^{n-1} < c_1 < c_2 < 2^n$, such that $c_1 h(x) \leq h(2x) \leq c_2 h(x)$, for all small x.

(C3). Since $\lim_{x \to 0+} h(x)/x^{n-1} = 0$, there exists some constant $c > 0$ such that $h(x) \leq cx^{n-1}$ for all small x.

(C4). Finally, it is immediate that $\lim_{x \to \infty} h(x)/x^n = 0$.

Example 4 We can also show that (still for $n \geq 1$)

$$h(x) = \frac{x^d}{\left(\ln(\ln(\frac{1}{x} + 1) + 1)\right)^a},$$

satisfies (C1)—(C4) for any $d \in (n - 1, n)$ and $a \geq 0$.

(C1). It is clear that $h : (0, \infty) \to (0, \infty)$ is a strictly increasing function and $\lim_{x \to 0+} h(x) = 0$, $\lim_{x \to \infty} h(x) = \infty$.

(C2). It was shown in Example 2 (H2) that $\lim_{x \to 0+} h(2x)/h(x) = 2^d$. Thus there exist constants c_1, c_2, with $2^{n-1} < c_1 < c_2 < 2^n$, such that $c_1 h(x) \leq h(2x) \leq c_2 h(x)$.

(C3). Since $\lim_{x \to 0+} h(x)/x^{n-1} = 0$, there exists some constant $c > 0$ such that $h(x) \leq cx^{n-1}$ for all small x.

(C4). The last assertion $\lim_{x \to \infty} h(x)/x^n = 0$ can also be easily checked.

More generally, we can show similarly that the gauge functions involving iterated logarithms of the form

$$h(x) = \frac{x^d}{\underbrace{\left(\ln(\ln(\cdots \ln(\frac{1}{x} + 1) \cdots + 1) + 1)\right)^a}_{K}}$$

also satisfy (C1)—(C4) for any $d \in (n - 1, n), a \geq 0$ and $K \in \mathbf{N}$.

References

[Be1] M. V. Berry, *Distribution of modes in fractal resonators*, in: Structural Stability in Physics (W. Göttinger and H. Eikemeier, eds.), Springer-Verlag, Berlin, 1979, pp. 51-53.

[Be2] _____, *Some geometric aspects of wave motion: wavefront dislocations, diffraction catastrophes, diffractals*, in: Geometry of the Laplace Operator, Proc. Symp. Pure Math., Vol. 36, Amer. Math. Soc., Providence, R. I., 1980, pp. 13-38.

[BiSo] M. S. Birman and M. Z. Solomjak, *Spectral asymptotics of nonsmooth elliptic operators*, I & II, Trans. Moscow Math. Soc. **27** (1972), 3-50 & **28** (1973), 3-34.

[Bo] G. Bouligand, *Ensembles impropes et nombre dimensionnel*, Bull. Sci. Math. (2) **52** (1928), 320-344 & 361-376.

[BrCa] J. Brossard and R. Carmona, *Can one hear the dimension of a fractal?* Comm. Math. Phys. **104** (1986), 103-122.

[Ce1] A. M. Caetano, *Some domains where the eigenvalues of the Dirichlet Laplacian have non-power second term asymptotic estimates*, J. London Math. Soc. (2) **43** (1991), 431-450.

[Ce2] _____, *On the search for the asymptotic behavior of the eigenvalues of the Dirichlet Laplacian for bounded irregular domains*, Internat. J. Scientific Computing & Modelling (to appear).

[CoHi] R. Courant and D. Hilbert, *Methods of mathematical physics*, Vol. I, English transl., Interscience, New York, 1953.

[EvHa] W. D. Evans and D. J. Harris, *Fractals, trees and the Neumann Laplacian*, Math. Ann. **296** (1993), 493-527.

[Fa1] K. Falconer, *The geometry of fractal sets*, Cambridge Univ. Press, Cambridge, 1985.

[Fa2] _____, *Fractal geometry: mathematical foundations and applications*, Wiley, Chichester, 1990.

[Fa3] _____, *On the Minkowski measurability of fractals*, Proc. Amer. Math. Soc. (to appear).

[FlVa] J. Fleckinger-Pellé and D. G. Vassiliev, *An example of a two-term asymptotics for the "counting function" of a fractal drum*, Trans. Amer. Math. Soc. **337** (1993), 99-116.

[HeLa] C. Q. He and M. L. Lapidus, *Generalized Minkowski content and the vibrations of fractal drums and strings*, Mathematical Research Letters **3** (1996), 1-10.

[Jo] P. W. Jones, *Quasiconformal mappings and extendability of functions in Sobolev spaces*, Acta Math. **147** (1981), 71-88.

[KaSa] J.-P. Kahane and R. Salem, *Ensembles parfaits et séries trigonométriques*, Hermann, Paris, 1963.

[La1] M. L. Lapidus, *Fractal drum, inverse spectral problems for elliptic operators and a partial resolution of the Weyl-Berry conjecture*, Trans. Amer. Math. Soc. **325** (1991), 465-529.

[La2] _____, *Can one hear the shape of a fractal drum? Partial resolution of the Weyl-Berry conjecture*, in: Geometric Analysis and Computer Graphics (P. Concus *et al.*, eds.), Proc. MSRI Workshop (Berkeley, May 1988), Mathematical Sciences Research Institute Publications, Vol. 17, Springer-Verlag, New York, 1991, pp. 119-126.

[La3] _____, *Spectral and fractal geometry: from the Weyl-Berry conjecture for the vibrations of fractal drums to the Riemann zeta-function*, in: Differential Equations and Mathematical Physics (C. Bennewitz, ed.), Proc. Fourth UAB Intern. Conf. (Birmingham, March 1990), Academic Press, New York, 1992, pp. 151-182.

[La4] _____, *Vibrations of fractal drums, the Riemann hypothesis, waves in fractal media, and the Weyl-Berry conjecture*, in: Ordinary and Partial Differential Equations (B. D. Sleeman and R. J. Jarvis, eds.), Vol. IV, Proc. Twelfth Dundee Intern. Conf. (Dundee, Scotland, UK, June 1992), Pitman Research Notes in Mathematics Series **289**, Longman Scientific and Technical, London, 1993, pp. 126-209.

[LaFl] M. L. Lapidus and J. Fleckinger-Pellé, *Tambour fractal: vers une résolution de la conjecture de Weyl-Berry pour les valeurs propres du laplacien*, C. R. Acad. Sci. Paris Sér. I Math **306** (1988), 171-175.

[LaMa1] M. L. Lapidus and H. Maier, *Hypothèse de Riemann, cordes fractales vibrantes et conjecture de Weyl-Berry modifiée*, C. R. Acad. Sci. Paris Sér. I Math **313** (1991), 19-24.

[LaMa2] ____, *The Riemann hypothesis and inverse spectral problems for fractal strings*, J. London Math. Soc. (2) **52** (1995), 15-34.

[LaPo1] M. L. Lapidus and C. Pomerance, *Fonction zêta de Riemann et conjecture de Weyl-Berry pour les tambours fractals*, C. R. Acad. Sci. Paris Sér. I Math. **310** (1990), 343-348.

[LaPo2] ____, *The Riemann zeta-function and the one-dimensional Weyl-Berry conjecture for fractal drums*, Proc. London. Math. Soc. (3) **66** (1993), 41-69.

[LaPo3] ____, *Counterexamples to the modified Weyl-Berry conjecture on fractal drums*, Math. Proc. Cambridge Philos. Soc. **119** (1996), 167-178.

[M] B. B. Mandelbrot, *The fractal geometry of nature*, rev. and enl. ed., Freeman, New York, 1983.

[Ma] V. G. Maz'ja, *Sobolev spaces*, Springer-Verlag, Berlin, 1985.

[Me] G. Métivier, *Valeurs propres de problèmes aux limites elliptiques irréguliers*, Bull. Soc. Math. France, Mém. **51-52** (1977), 125-219.

[Od] A. M. Odlyzko, *The 10^{20}-th zero of the Riemann zeta-function and 175 millions of its neighbors*, preprint, AT&T Bell Labs, Murray Hill, 1991; and book to appear.

[ReSi] M. Reed and B. Simon, *Methods of modern mathematical physics*, Vol. IV, *Analysis of operators*, Academic Press, New York, 1978.

[Ro] C. A. Rogers, *Hausdorff measures*, Cambridge, 1970.

[Ti] E. C. Titchmarsh, *The theory of the Riemann zeta-function*, 2nd ed. (revised by D. R. Heath-Brown), Oxford University Press, 1986.

[Tr] C. Tricot, *Two definitions of fractal dimension*, Math. Proc. Cambridge Philos. Soc. **91** (1988), 57-74.

[We1] H. Weyl, *Das asymptotische Verteilungsgezetz der Eigenwerte linearer patieller Differentialgleichungen*, Math. Ann. **71** (1912), 441-479.

[We2] ____, *Über die Abhängigkeit der Eigenschwingungen einer Membran von deren Begrenzung*, J. Angew. Math. **141** (1912), 1-11.

CHRISTINA Q. HE   and   MICHEL L. LAPIDUS

Department of Mathematics
University of California
Sproul Hall
Riverside, CA 92521-0135, USA

e-mail address: che@math.ucr.edu, lapidus@math.ucr.edu

Editorial Information

To be published in the *Memoirs*, a paper must be correct, new, nontrivial, and significant. Further, it must be well written and of interest to a substantial number of mathematicians. Piecemeal results, such as an inconclusive step toward an unproved major theorem or a minor variation on a known result, are in general not acceptable for publication. *Transactions* Editors shall solicit and encourage publication of worthy papers. Papers appearing in *Memoirs* are generally longer than those appearing in *Transactions* with which it shares an editorial committee.

As of January 31, 1997, the backlog for this journal was approximately 7 volumes. This estimate is the result of dividing the number of manuscripts for this journal in the Providence office that have not yet gone to the printer on the above date by the average number of monographs per volume over the previous twelve months, reduced by the number of issues published in four months (the time necessary for preparing an issue for the printer). (There are 6 volumes per year, each containing at least 4 numbers.)

A Copyright Transfer Agreement is required before a paper will be published in this journal. By submitting a paper to this journal, authors certify that the manuscript has not been submitted to nor is it under consideration for publication by another journal, conference proceedings, or similar publication.

Information for Authors and Editors

Memoirs are printed by photo-offset from camera copy fully prepared by the author. This means that the finished book will look exactly like the copy submitted.

The paper must contain a *descriptive title* and an *abstract* that summarizes the article in language suitable for workers in the general field (algebra, analysis, etc.). The *descriptive title* should be short, but informative; useless or vague phrases such as "some remarks about" or "concerning" should be avoided. The *abstract* should be at least one complete sentence, and at most 300 words. Included with the footnotes to the paper, there should be the 1991 *Mathematics Subject Classification* representing the primary and secondary subjects of the article. This may be followed by a list of *key words and phrases* describing the subject matter of the article and taken from it. A list of the numbers may be found in the annual index of *Mathematical Reviews*, published with the December issue starting in 1990, as well as from the electronic service e-MATH [**telnet e-MATH.ams.org** (or **telnet 130.44.1.100**). Login and password are **e-math**]. For journal abbreviations used in bibliographies, see the list of serials in the latest *Mathematical Reviews* annual index. When the manuscript is submitted, authors should supply the editor with electronic addresses if available. These will be printed after the postal address at the end of each article.

Electronically prepared papers. The AMS encourages submission of electronically prepared papers in $\mathcal{A}_{\mathcal{M}}\mathcal{S}$-TeX or $\mathcal{A}_{\mathcal{M}}\mathcal{S}$-LaTeX. The Society has prepared author packages for each AMS publication. Author packages include instructions for preparing electronic papers, the *AMS Author Handbook*, samples, and a style file that generates the particular design specifications of that publication series for both $\mathcal{A}_{\mathcal{M}}\mathcal{S}$-TeX and $\mathcal{A}_{\mathcal{M}}\mathcal{S}$-LaTeX.

Authors with FTP access may retrieve an author package from the Society's internet node **e-MATH.ams.org** (130.44.1.100). For those without FTP access, the author package can be obtained free of charge by sending e-mail to `pub@math.ams.org` (Internet) or from the Publication Division, American Mathematical Society, P.O. Box 6248, Providence, RI 02940-6248. When requesting an author package, please specify $\mathcal{AMS}$-TEX or $\mathcal{AMS}$-LATEX, Macintosh or IBM (3.5) format, and the publication in which your paper will appear. Please be sure to include your complete mailing address.

Submission of electronic files. At the time of submission, the source file(s) should be sent to the Providence office (this includes any TEX source file, any graphics files, and the DVI or PostScript file).

Before sending the source file, be sure you have proofread your paper carefully. The files you send must be the EXACT files used to generate the proof copy that was accepted for publication. For all publications, authors are required to send a printed copy of their paper, which exactly matches the copy approved for publication, along with any graphics that will appear in the paper.

TEX files may be submitted by email, FTP, or on diskette. The DVI file(s) and PostScript files should be submitted only by FTP or on diskette unless they are encoded properly to submit through e-mail. (DVI files are binary and PostScript files tend to be very large.)

Files sent by electronic mail should be addressed to the Internet address `pub-submit@math.ams.org`. The subject line of the message should include the publication code to identify it as a Memoir. TEX source files, DVI files, and PostScript files can be transferred over the Internet by FTP to the Internet node `e-math.ams.org` (130.44.1.100).

Electronic graphics. Figures may be submitted to the AMS in an electronic format. The AMS recommends that graphics created electronically be saved in Encapsulated PostScript (EPS) format. This includes graphics originated via a graphics application as well as scanned photographs or other computer-generated images.

If the graphics package used does not support EPS output, the graphics file should be saved in one of the standard graphics formats—such as TIFF, PICT, GIF, etc.—rather than in an application-dependent format. Graphics files submitted in an application-dependent format are not likely to be used. No matter what method was used to produce the graphic, it is necessary to provide a paper copy to the AMS.

Authors using graphics packages for the creation of electronic art should also avoid the use of any lines thinner than 0.5 points in width. Many graphics packages allow the user to specify a "hairline" for a very thin line. Hairlines often look acceptable when proofed on a typical laser printer. However, when produced on a high-resolution laser imagesetter, hairlines become nearly invisible and will be lost entirely in the final printing process.

Screens should be set to values between 15% and 85%. Screens which fall outside of this range are too light or too dark to print correctly.

Any inquiries concerning a paper that has been accepted for publication should be sent directly to the Editorial Department, American Mathematical Society, P. O. Box 6248, Providence, RI 02940-6248.

Selected Titles in This Series

MS catalog for earlier tit